LEARN MATH FAST SYSTEM
VOLUME IV, SECOND EDITION

By J. K. Mergens

Learn Math Fast System Vol. IV Second Edition
Copyright © 2011 and ©2021 Registration Number TX 7-316-060
ISBN: 978-1523636006
www.LearnMathFastBooks.com

Image of United States coin from United States Mint

CONTENTS

INTRODUCTION

Welcome to the *Learn Math Fast System, Volume IV*. This book covers basic Geometry. For best results, you should read Volumes I - III of the *Learn Math Fast System*, first. But if you are already proficient in basic math and pre-algebra, then you are ready for this book.

Be sure to read each lesson and then complete the worksheet or test at the end of the lesson. Compare your answers with the ones in the back of the book. If you get stuck on a problem, use the answer key to help you solve it.

This book comes with a Geometry Kit, which can be purchased separately on our website LearnMathFastBooks.com. This kit has some handy tools to help you understand and remember geometric concepts. As you read through the book, you will learn about each item in the kit.

Once you finish this book, you are ready for Volume V of the *Learn Math Fast* series.

CHAPTER 1

LINES AND ANGLES

LESSON 1: LINES

Geometry is the study of shapes, angles, and lines. In geometry, a *line* goes on forever with no end. Since we can't draw a line that goes on forever, we put arrows on the end to show that it does. The line below is called, "line AD." In geometry, it is written $\overleftrightarrow{AD}$ for short. Notice that the little symbol on top of "AD," in the last sentence, is a picture of a miniature line. You can tell that the drawing below is a "line" because it has arrows on both ends to show that it goes on forever.

Since lines are so long, we usually only talk about one little chunk of a line at a time. In geometry, when you talk about just one small chunk of a line, you mark it with two points. These points are called *endpoints* and they create a *line segment*. Below is line AD with some points drawn on it. These points, B and C, show just a small chunk of the line.

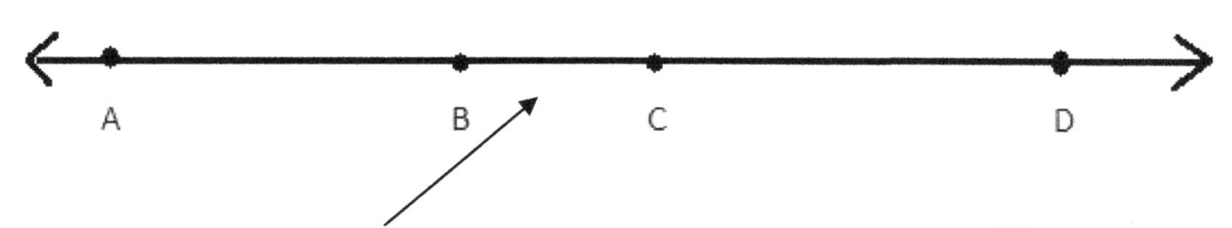

This is called *line segment* BC. To make it shorter, we write $\overline{BC}$ to say line segment BC. Again, the symbol above "BC" is just a miniature line segment.

Remember this: a line has an arrow on both ends. A line segment has an endpoint on both ends.

Below, I've drawn another type of line; it is called a *ray*. A *ray* is different from a line or a line segment because it has ONE endpoint and ONE arrow. The arrow goes on forever, but the end has a fixed point called the *endpoint*.

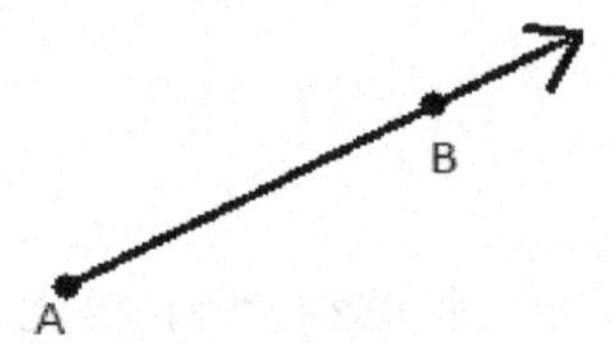

Think of a ray as a "sun ray." The endpoint is attached to the sun and the arrow is the sun RAY shooting out to the earth.

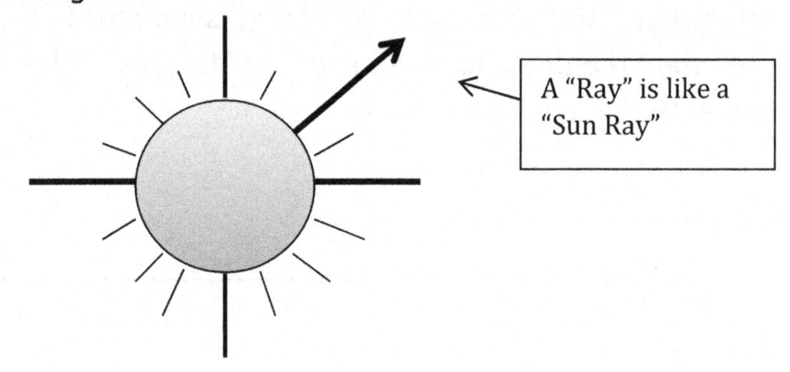

The point that is attached to the sun is called the *endpoint*. The endpoint in ray AB below is called endpoint A.

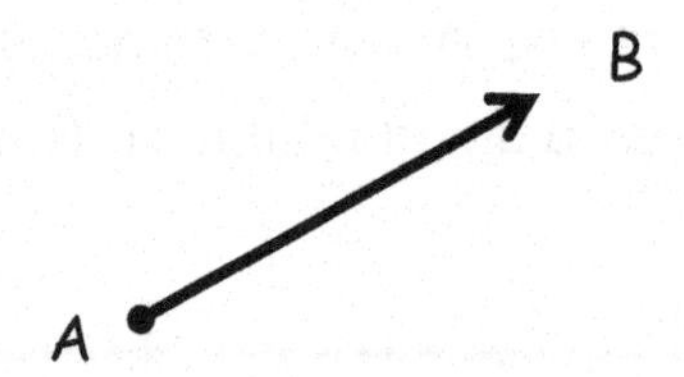

Ray AB is written $\overrightarrow{AB}$ for short. Notice the symbol above AB. It is a picture of a miniature ray.

When two rays join together at their endpoints, they can't help but create an *angle.*

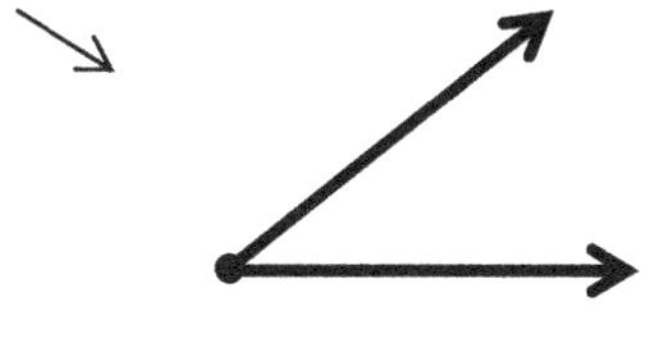

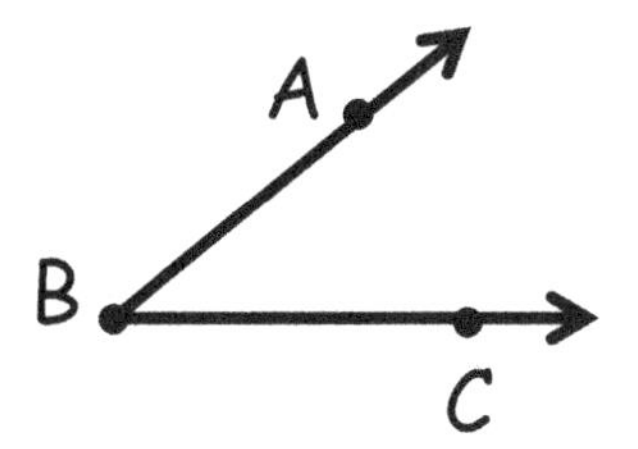

The picture above is an *angle.* It is written $\angle ABC$. That is the long name. You can call it angle B for short because that is the letter at the endpoint. When endpoints come together and make an angle, that new combined endpoint is called a *vertex.* Can you name the vertex in the angle above? The vertex of the angle is B.

Let's go over the new geometry words you learned in this lesson. First, you learned that a line goes on forever, so it is drawn with an arrow on both ends. A small piece of that line in called a line segment and it has an endpoint on both ends. A ray has one of each, an arrow on one end and an endpoint on the other. When two rays join endpoints, they morph and become an angle with a vertex.

Take a quick test to make sure you understand everything so far. Be sure to include arrows in your drawings whenever necessary.

Name: ______________________________ Date: ______________

WORKSHEET 4-1

1. Draw a line with two points creating a line segment AB.

2. Draw an angle and label the points as CDE with point D being the vertex.

3. Draw ray KL with K as the endpoint.

4. Name each picture as a line, a line segment, a ray, or an angle.

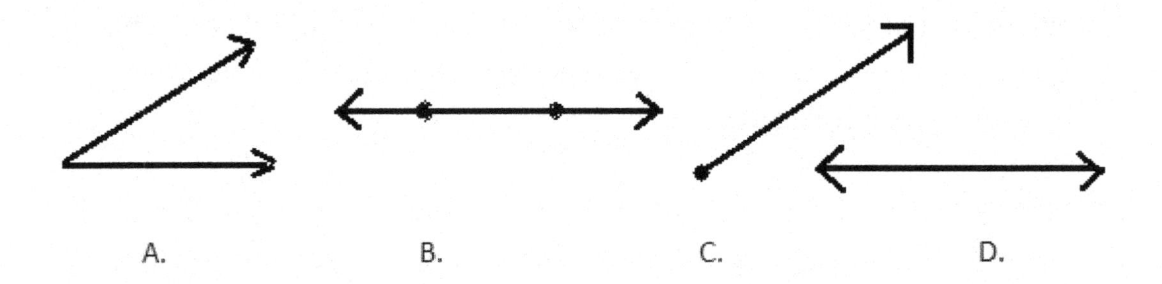

A. B. C. D.

5. Using the little symbols we just learned: ⟷, ⟶ , —, and ∠, write the short name for each of the following. For example, Line EF would be written as $\overleftrightarrow{EF}$.

   a. Line AB
   b. Line segment CD
   c. Ray BC
   d. Angle EFG

LESSON 2: ANGLES

If you correctly answered all the problems on the last worksheet, then you are ready to move on. If you got a wrong answer, go back, find out why and then move on.

You will need the protractor from the Geometry Kit for this next lesson. If you don't have the Geometry Kit, any other protractor will work. Look at your protractor. The numbers go around the edge from 0 to 180 in both directions. But no matter which direction you go, 90 will always be on top, in the center.

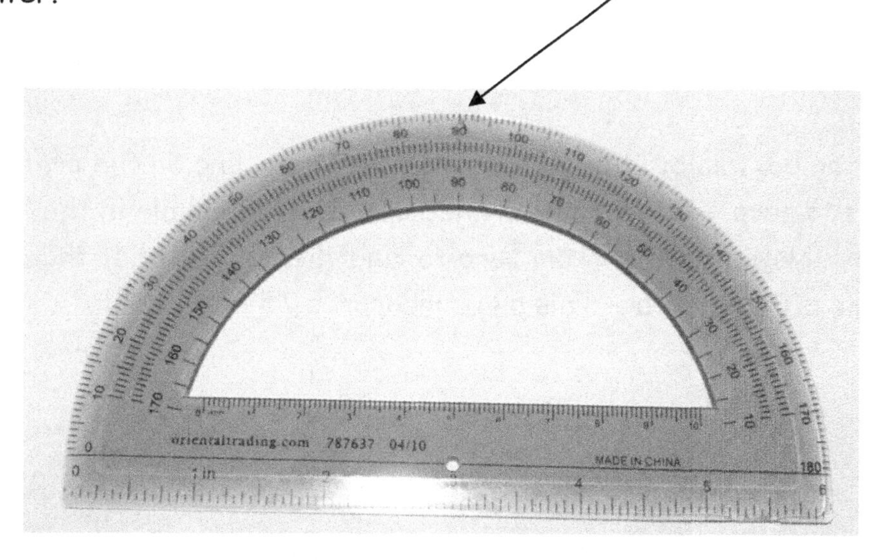

A protractor is a tool that we use to measure an angle. Angles are measured in *degrees*. Let me explain. Below is a circle with an angle drawn inside of it. The entire circle measures 360 degrees. Our angle below is only 75 of the 360 degrees. A protractor is used to find out exactly how many degrees this angle measures.

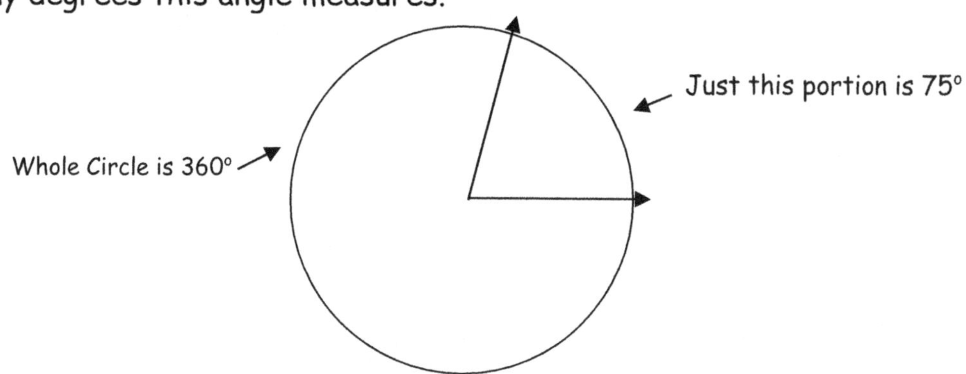

Look at the picture to see how a protractor can measure this angle.

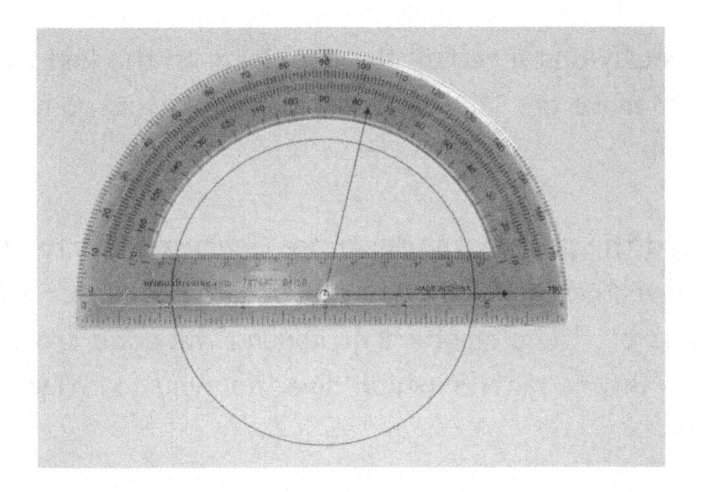

I lined up the bottom of the angle with the straight line on the protractor that points to zero. The vertex is lined up with the pinhole in the protractor. Now, count up from zero to find the number that the other ray of the angle is pointing to. This angle measures 75 degrees.

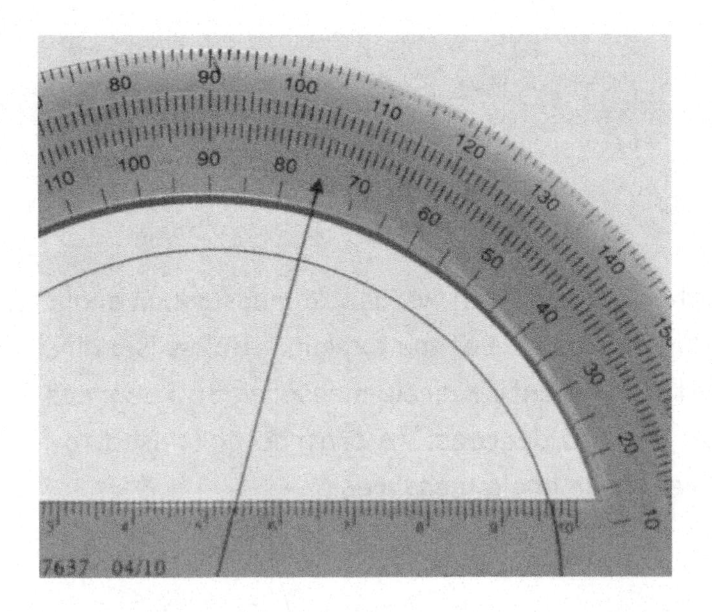

Our protractor is only half of a circle. It measures from 1 to 180 degrees.

Below is a picture of an angle that is so big, it is almost the entire half circle. It measures about 179°. It is nearly a straight line.

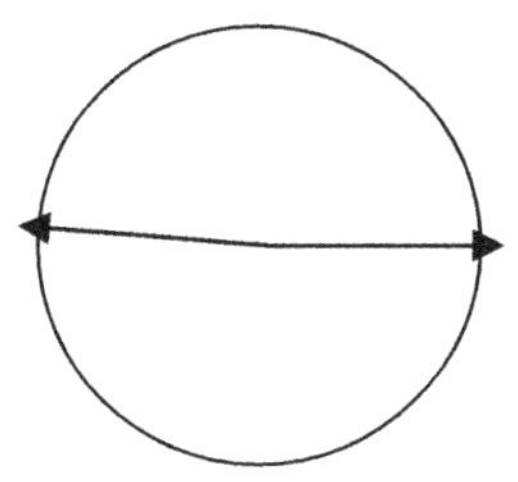

If this angle gets 1 degree bigger, it WILL BE a straight line; also called a straight angle. It is hard to tell the difference between a straight line and a straight angle, but technically a straight angle would be 180 degrees and a line...is just a line.

You may have heard someone say, "He made a one-eighty and came right back." Look at your protractor. A completely straight angle, which looks just like a straight line, is 180 degrees. So, if someone were walking with their arm pointing straight out in front of them and then turned around to face the opposite direction, their arm would have made a 180 degree arc; that's why making a "180" is making a complete turn-around.

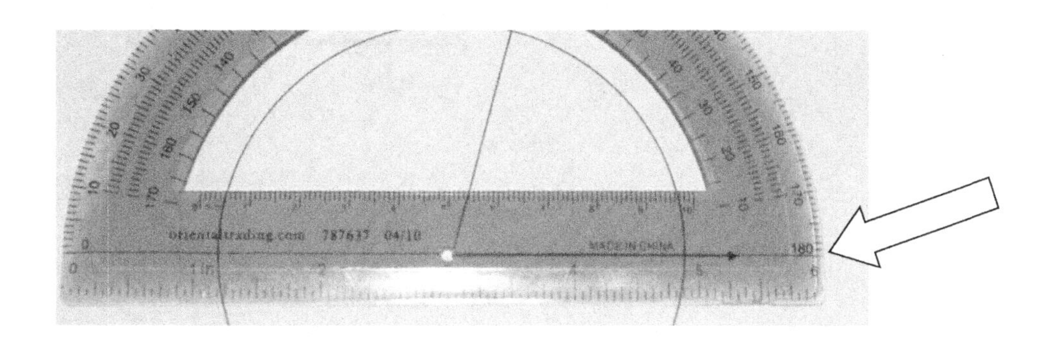

Instead of writing out the word "degrees" every time, you can use a symbol. The symbol for degrees is a little circle drawn towards the top, right side of a number like this, 180°.

Often times, during a sports competition you'll hear, "He did a three-sixty!" When someone does a "360" on a snowboard, it means he jumped up and made a complete circle before landing. That's because turning in a complete circle is a 360° turn. Look at your protractor again. If you start at zero and spin a half circle, it is 180°. If you keep spinning to make a full circle, you will have gone twice that much or 360 degrees. That's why the symbol for degrees is a little circle above the number. At least I think that's why - it makes sense.

360°

We use a protractor to measure angles in degrees. Below, I have drawn a perfect 90° angle. Find the vertex of this angle.

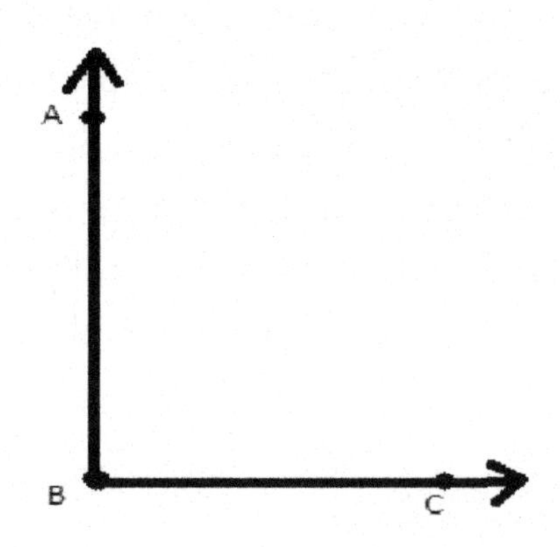

Now look at your protractor again. Along the flat edge in the very center you will see a tiny hole.

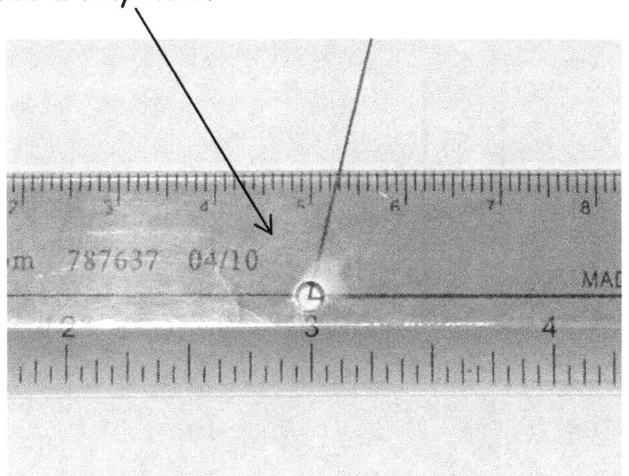

Put that over top of the vertex and line up ray BC with the 0° line drawn on the protractor.

Once you have the vertex lined up with the center hole and the "zero" lines are lined up together, you will see that ray BA is pointing to exactly 90°. That's how you know it is a 90° angle. A perfect 90° angle is called a *right angle.*

A lot of math books draw a little box in the corner of a right angle to prove that the angle is exactly 90°.

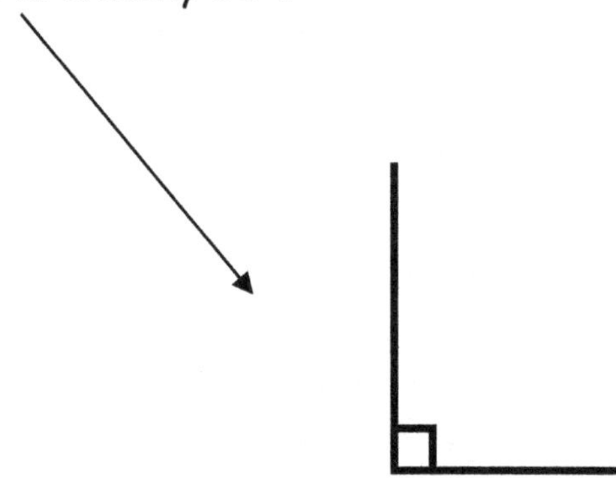

This little box tells us that this angle is not 89°, and it is not 91°, it is exactly 90°. It is also called a right angle. Knowing it is exactly 90° will be helpful later in geometry when we have to try to solve mystery numbers. Below are two 90° angles, back to back.

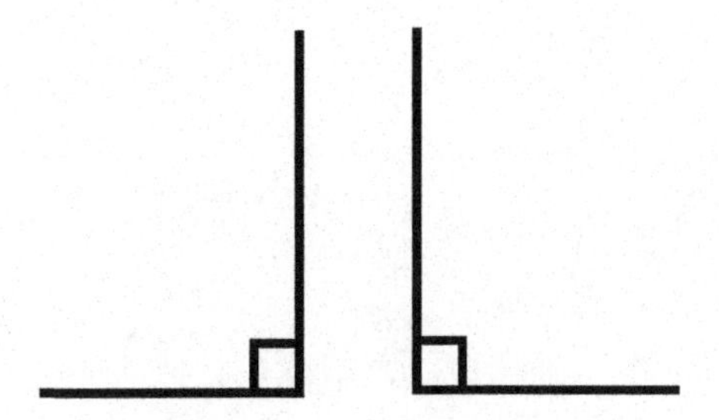

Now let's squish them together.

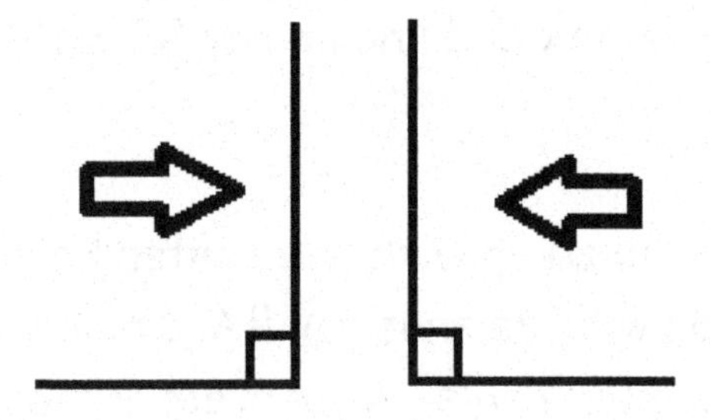

Now they look like this…

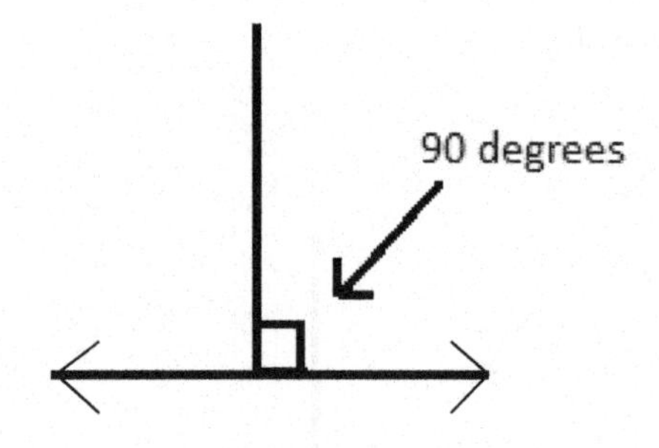

Now we have our first mystery. It is a super easy mystery to solve, but it is still a mystery. What is the measurement of the angle without a box?

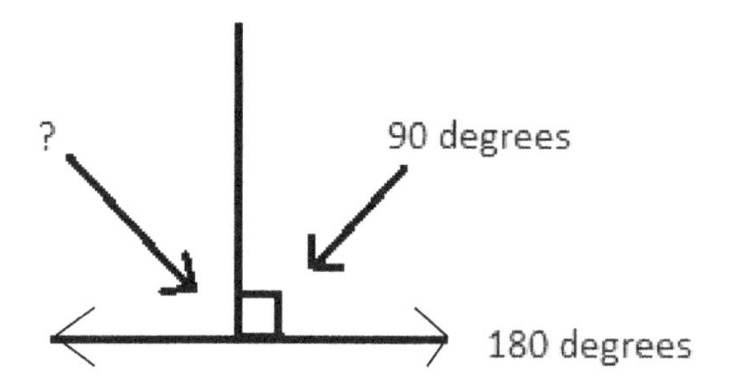

That's simple; you know the other angle is 90° because we just squished two 90° angles together. But you could have figured it out mathematically. Earlier we learned that a straight line is 180°, so whatever angles are on top have to equal 180°. If one side is 90°, then the other side must also be 90° because 90 + 90 = 180.

The majority of the geometry questions you will answer will be "find the missing number" type of questions. You will be given a few clues and you use those clues to figure out the missing number.

Now what do you suppose would happen if I moved this line over by 1°?

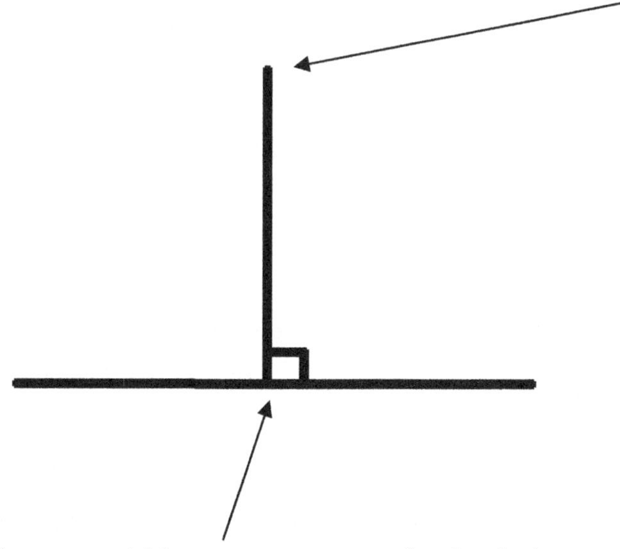

First of all, we would have to remove the little box in the corner because it would no longer be 90°.

Also, if I made that angle 1° bigger, it would become a 91° angle.

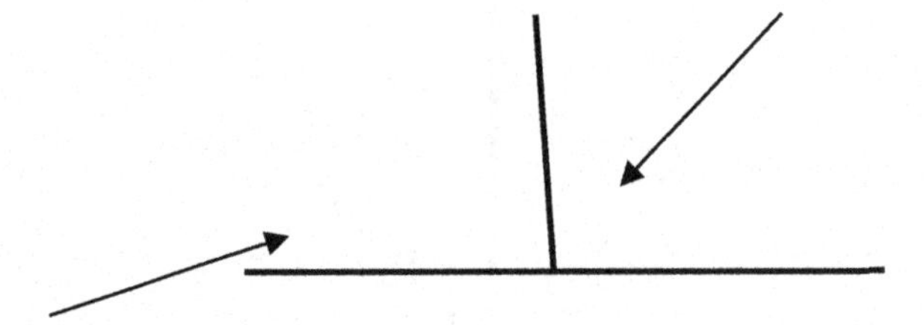

Now how big is this angle?

When I moved the line over by 1 degree, it made the angle bigger. But the other angle got 1 degree smaller, so it must be 89° now. You could also figure this out mathematically, instead of logically. You know the straight line measures 180°. You know one angle measures 91°. Do the math 180 - 91 = 89. That proves the angle is 89°.

Now I will move the other line even further. What is the missing angle?

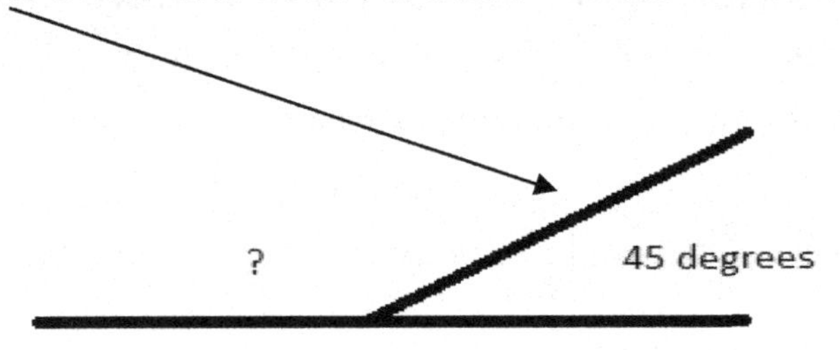

Since a straight line is 180°, we know that the two angles must equal 180. We know one of the angles is 45°, so the other angle must be 135°.

If this seems super easy, you're right, it is. If this sounds complicated, you are over thinking it. Go back and read the last few pages again. This should be as easy as adding numbers together that equal 180.

Earlier I told you that a 90° angle is called a right angle. Well, here is another geometry fact. Any angle that is smaller than 90° is called an *acute* angle. You can remember this by thinking, "Oh look at that *cute* little angle."

Any angle that is bigger than 90° is called an *obtuse* angle. The 135° angle from the last problem is an obtuse angle and the 45° angle is an acute angle.

These two words are very important in geometry, so you have to learn them. If you don't know the difference between an acute angle and an obtuse angle, then you don't know geometry. Look at the angles below. Which ones are acute? Which ones are obtuse? Are any of them a right angle?

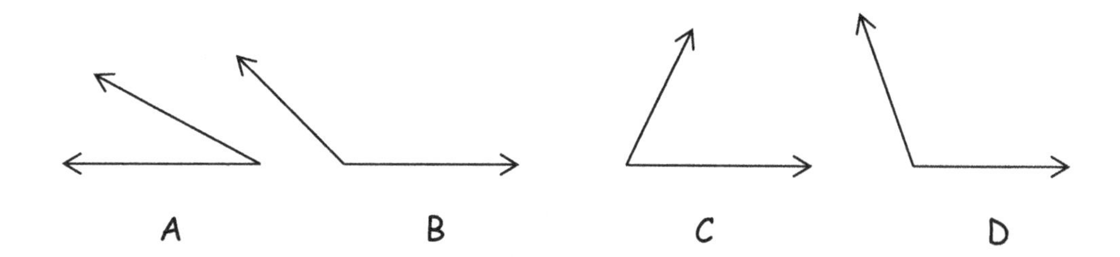

A right angle is a perfect 90° angle, so none of them are a right angle. Angles A and C are smaller than 90°, so they are acute. Angles B and D are larger than 90°, so they are obtuse.

When two angles share one side and share a vertex, they are called *adjacent angles.* Look at the drawing below.

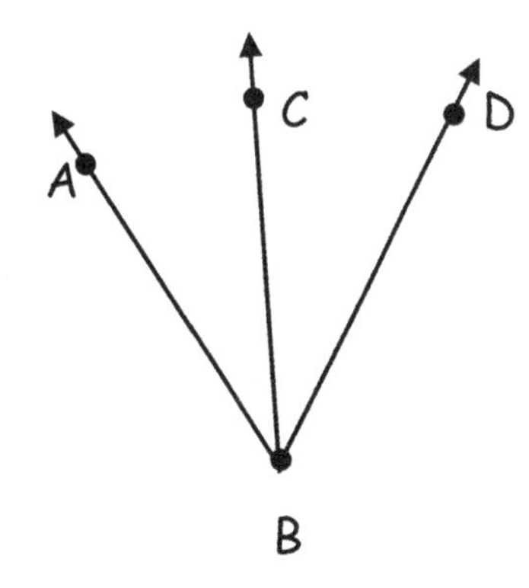

Angles ABC and CBD are adjacent angles because they share a side, BC, and they share vertex B.

Let's review those last three new words: acute, obtuse, and adjacent angles. An acute angle is smaller than 90°. It's "a cute" little angle. An obtuse angle is bigger than 90°- it's obese. And adjacent angles are right next to each other sharing one side and a vertex.

Name: ______________________ Date: ______________

WORKSHEET 4-2

Look at the drawing below and then answer the questions.

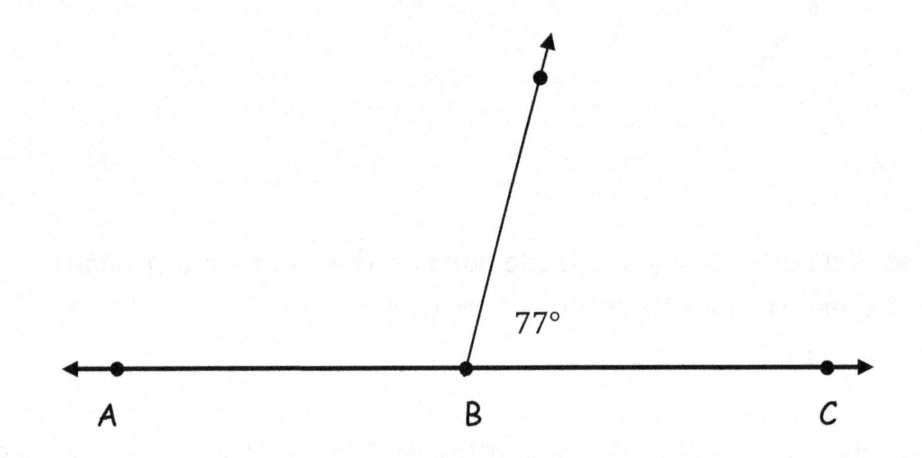

1. Is angle ABD obtuse or acute?

2. Is angle DBC obtuse or acute?

3. What is the measurement of angle ABD?

4. What is the measurement of angle DBC?

5. Look at angle ABD. Which letter is the vertex?

6. Name the angle that is adjacent to angle DBC.

LESSON 3: OPPOSITE ANGLES

If that last worksheet was easy and you got all the answers correct, keep reading. If you had any problems, go back and learn what you missed. Look at this next drawing. The only clue you are given is that one angle is 45°. That is all we need to figure out the rest of the angles. We know this angle must be 135° because 180 - 45 = 135.

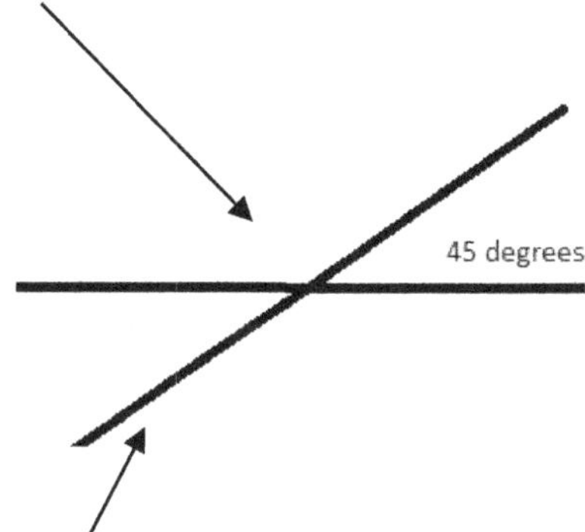

But wait! This is a straight line too, so the angles on top of IT must also equal 180°. If one side of this line is 135°, then the other side must be 45°.

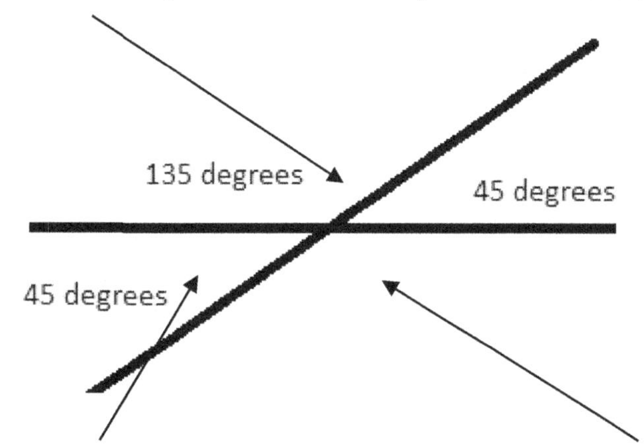

And if we know that this angle is 45°, then we know that this angle is 135°. In fact, *opposite angles* are always equal. The two 45° angles above are called *opposite angles*. The two 135° angles above are also opposite angles. Opposite angles share a vertex, but not a side. Look at these next two lines. They form a perfect cross.

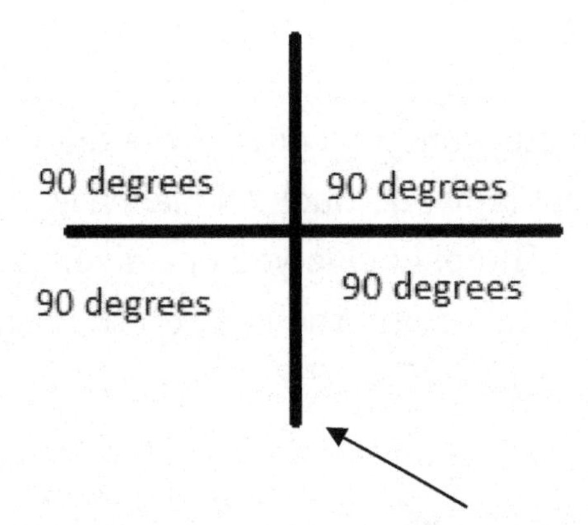

Can you guess what would happen, if I moved this line one way or the other?

I will move it over a little. Watch how the angles change, when I push the line over 20°.

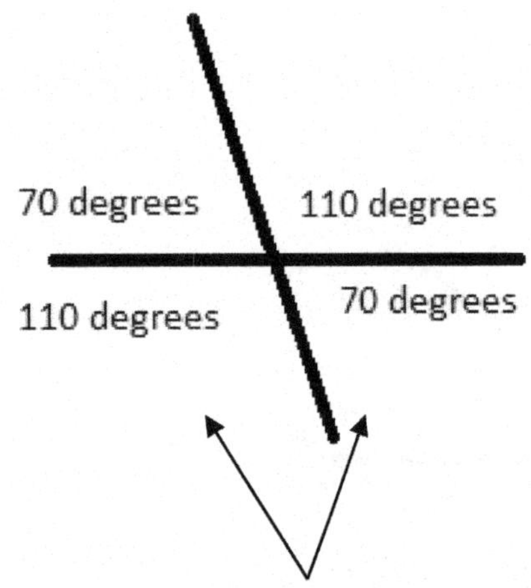

Notice how two angles on a straight line always equal 180 and how opposite angles are always the same. Those are two big clues to figuring out missing numbers. Fill in the correct size for angles A, B, and C below without using a protractor.

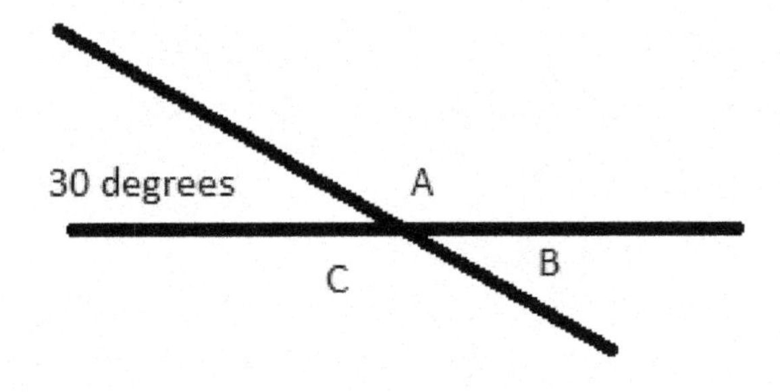

Our first clue is that opposite angles are equal. Angle B is opposite a 30° angle. Since opposite angles are always equal, angle B must be 30° too. And we know that when two angles form a straight line, they total 180°. To figure out angle A, we subtract 180 - 30 = 150. Angle A = 150°. Opposite angles are always equal, so angle C is also 150°. Here are the answers.

Angle A = 150°
Angle B = 30°
Angle C = 150°

Look at this next picture. What size is each angle?

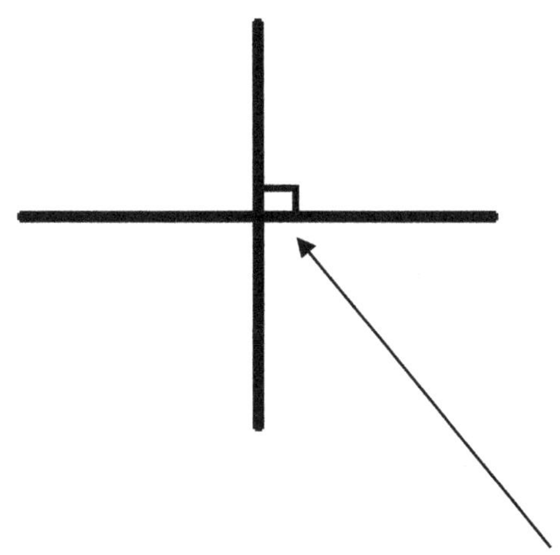

Do you remember what the little box means? It means this angle is exactly 90°. With that one little clue you should be able to figure out all 4 angles. Remember our clues: Opposite angles are always equal and angles that form a straight line always equal 180. Have you figured out the answers? That's right, all 4 angles are 90° each. Together they total 360°, just like a circle.

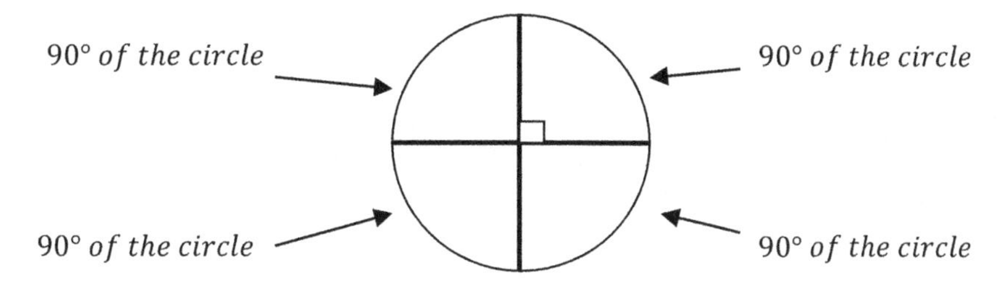

Take the Chapter Review Test, to see if you are ready to continue. You must get all the answers correct before you can continue.

Name: ______________________________ Date: ____________

CHAPTER 1 REVIEW TEST

All questions will be about this drawing:

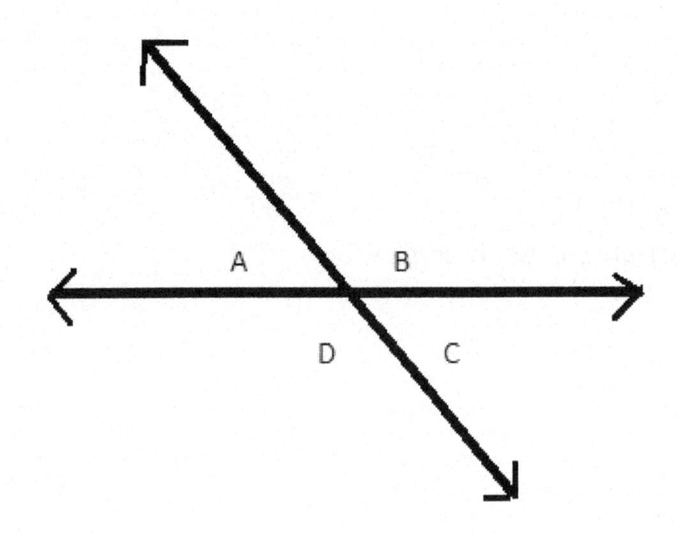

1. Is angle B an obtuse angle or an acute angle?
2. Is angle C an obtuse angle or an acute angle?
3. If angle A is 50°, what size are angles B, C, and D?
4. Are angles A and D opposite or adjacent angles?
5. Are angles A and B opposite or adjacent angles?
6. If angle D is 129°, what size is angle B?
7. If angle D is 129°, what size is angle C?
8. Why are there arrows on the ends of the lines?
9. Draw a picture to portray each of the following geometric terms.
   A Line segment
   A Ray
   An angle with a vertex Q
   Any Angle
   A Right angle
   An Obtuse angle
   An Acute angle
   Two opposite angles
   Two adjacent angles

CHAPTER 2

TRIANGLES, SQUARES, AND POLYGONS

LESSON 4: THE 3 SPECIAL TRIANGLES

If you had problems with that last worksheet, please go back and read that lesson again. You should understand all the new words and clues before attempting to learn more.

The first geometric shape we will learn about is the triangle. There are a few different kinds of triangles. We will start with learning what it takes for a shape to be called a triangle and then we will get into the different types.

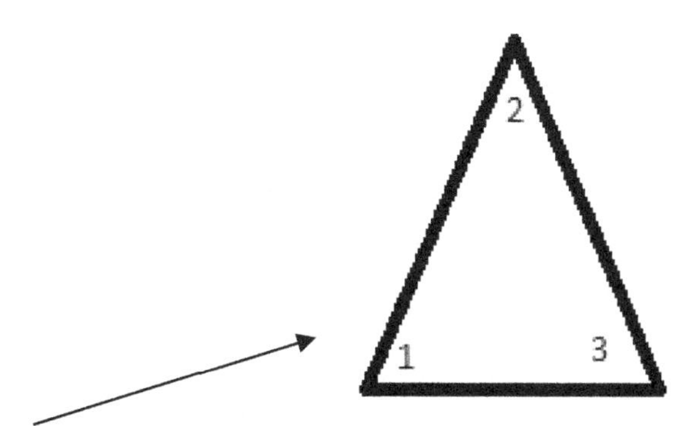

This is a *triangle*. A triangle has 3 angles. "Tri" means 3. That's why we call a tri-cycle a tricycle because it has 3 wheels. There are a lot of different ways to draw a triangle. Here are just a few.

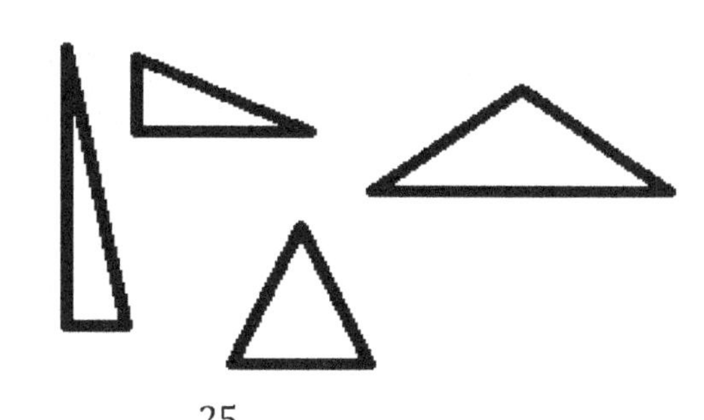

Since all those shapes have 3 angles, they are all called triangles.

Below is a picture of a right angle. I know it is a right angle because of the little box drawn in the corner.

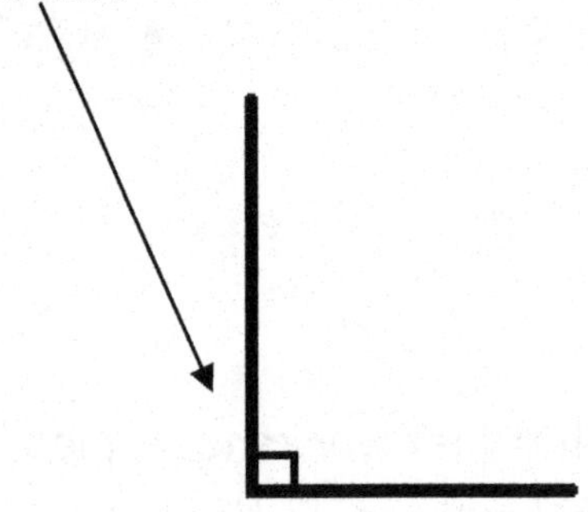

If I connect these two lines with a third line, it will form a triangle. This type of triangle is called a *right triangle.* Can you guess why? Of course you can.

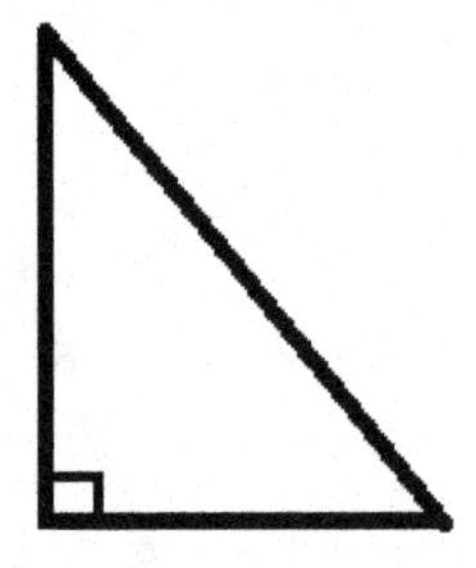

It is called a *right triangle* because one of the angles is exactly 90°- a right angle. Sometimes, a right triangle is drawn with the right angle on top.

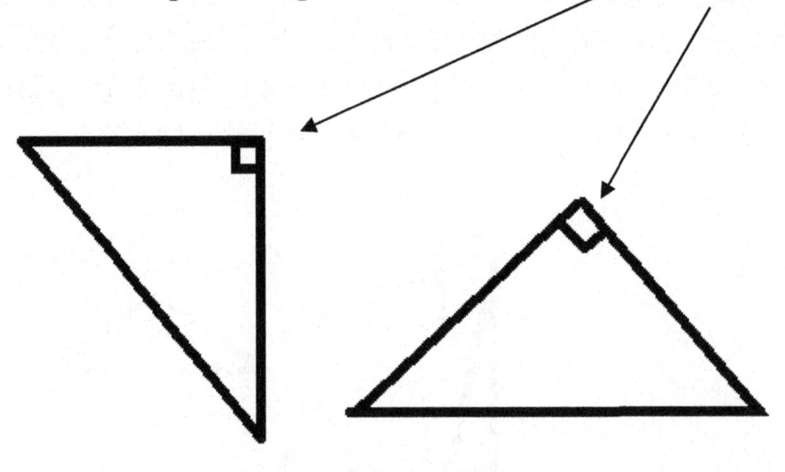

The first big clue we will learn about triangles is that all three angles will always total 180°; just like a straight line. So, if you know the size of at least two of the angles in a triangle, you can always figure out the third angle. For example, below is a RIGHT triangle, so automatically you already know one of the measurements; it's $90°$. Now let's say that this angle is $50°$. What is the measurement of the third angle?

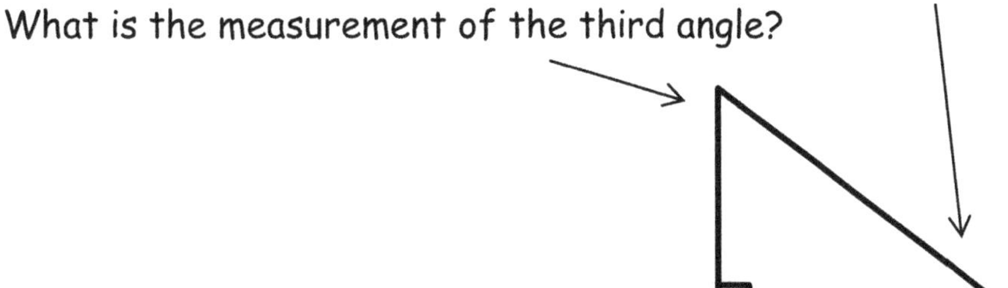

Let's do the math. All three angles HAVE TO equal $180°$. It has no choice because it is a triangle. I can actually figure out this one in my mind, but I'm going to write out an algebraic equation; just to be fancy.

$$90° + 50° + x° = 180°$$

I'll do the algebra for you below.

$$90° + 50° + x° = 180°$$
$$140° + x° = 180°$$
$$x° = 180° - 140°$$
$$x° = 40°$$

The third angle is $40°$. All triangles work that way, not just right triangles. If you know two of the angle's measurements, you can always figure out the third angle because all three HAVE TO total $180°$. The nice thing about working with Right triangles is that you automatically know one of the angle measurements, so all you need is one more.

The whole trick to geometry is learning all of these little "clues" so you can find a missing number. When a question says, "Look at the right triangle," it is giving you a secret clue that one of the angles is $90°$. Our job is to learn all of these "clues."

The next kind of triangle we will learn about is called an *isosceles* (eye-saw-sa-lease) triangle. It is different than a Right Triangle. When a triangle has two sides that are the same length, it is called an isosceles triangle.

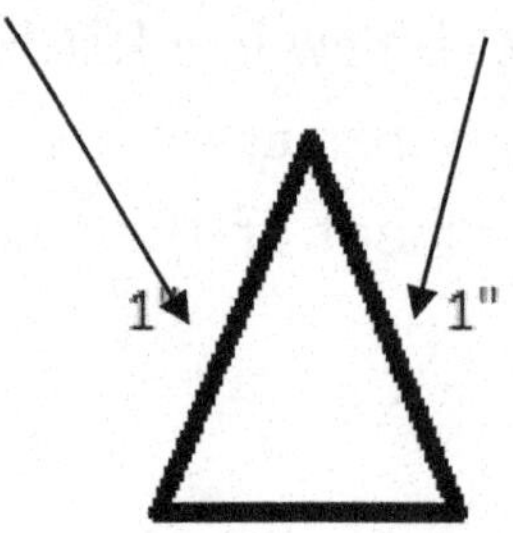

Get out a piece of paper and use the ruler on the edge of your protractor to draw an angle with two lines that are both 3" long. It doesn't matter what size of angle you draw, but the two lines MUST be the exact same length.

Now draw a third line to turn your angle into a triangle. Use your protractor to find the size of all three angles.

If you have drawn a true isosceles triangle, two of the angles will be exactly the same. If two of the angles in the triangle you drew are not equal, you need to double check that two of the sides are the same length. The two equal angles are called the *base angles.*

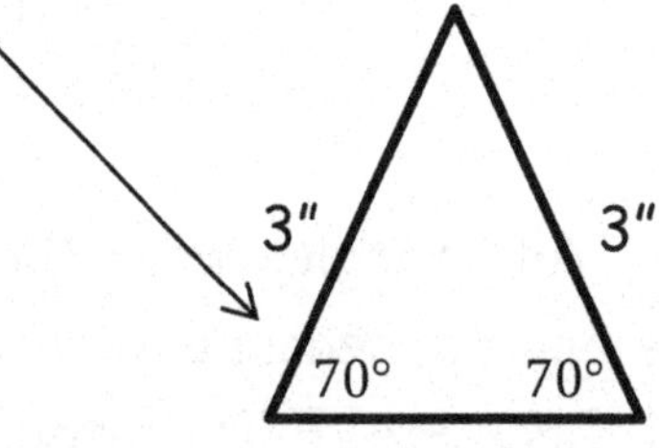

This little fact about an isosceles triangle is another clue to helping us find missing numbers. For example, look at the isosceles triangle below. This base angle measures 35°. What is the measurement of the other two angles?

Base Angle = 35°
Base Angle = ____°
Top Angle = ____°

Did you come up with 110° for the top angle? Good, then you know that the base angles of an isosceles triangle are equal and that all three angles HAVE TO equal 180°.

Draw three more isosceles triangles - all different sizes. No matter what the size, if two sides are the same length, then two base angles will be the same too.

Right triangles and Isosceles triangles are called *special triangles* because they have clues within them that help us figure out missing numbers. Look at this next triangle, for example.

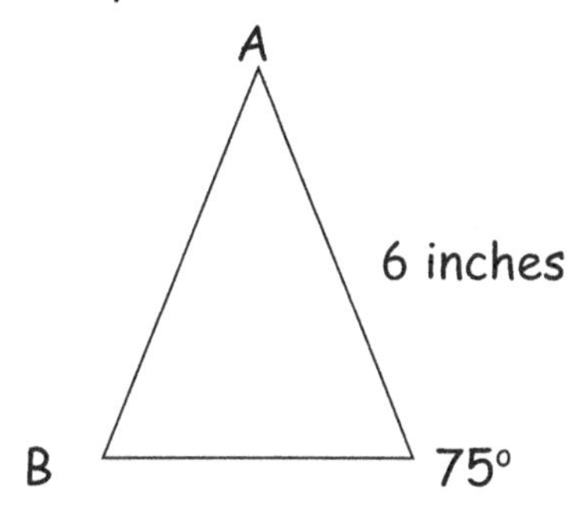

The triangle above is an isosceles triangle. Can you tell me the size of all 3 angles and at least 2 lengths? Try to figure it out on your own before you read on.

An isosceles triangle has at least two equal lengths, so we know that side AB is also 6". And if the two sides are the same length, then the two base angles have to be the same too, so Angle B must also be 75°. With a little math, we can figure out Angle A. 75 + 75 = 150. All 3 angles must equal 180, so 180 – 150 = 30. $\angle A = 30^\circ$.

Think about that triangle's name; isosceles. "I-saw-ce-lease two lengths were equal." Do you hear that? Isosceles...I-saw-at-least? Maybe that will help you remember what is so special about an isosceles triangle.

The third and final special triangle we will learn about is called an *Equilateral* triangle. Let's take a close look at that word. Equi-lateral. "Equi" means equal and "lateral" means side. All you really need to remember is that

"Equi" means equal because an equilateral triangle is completely equal. All three sides are equal and all three angles are equal. Equal, equal, equal!

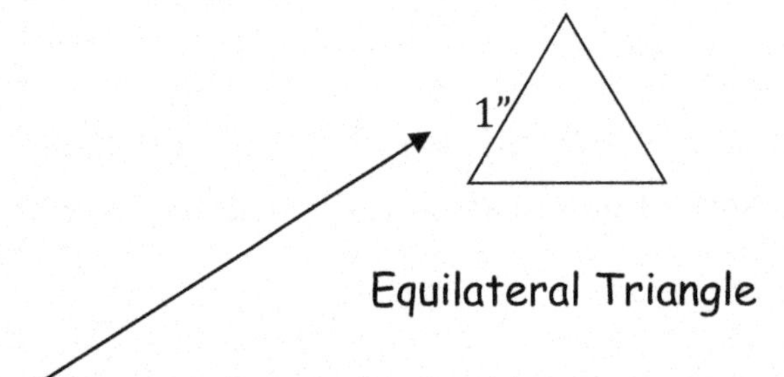

Equilateral Triangle

With this little clue we know that all three sides are each 1" long. And since all three angles are equal, they would have to be 60° each because 180 ÷ 3 = 60. It can't be any other way.

Let's review the three special triangles. Get out the triangle shapes from your Geometry Kit. First find the right triangle; the one with the 90° angle.

Next, find the isosceles triangle. It's the one with at least two equal sides. And of course, if you have 2 equal sides, you can't help but have 2 equal angles.

The third special triangle is the Equal, Equal, Equilateral. Everything is equal; the sides, the angles…it's perfect. And if all three angles are the same, they have to be 60° each. It's the only way.

Study each one of these triangles and memorize what is special about each one of them. When you are ready, complete the next worksheet.

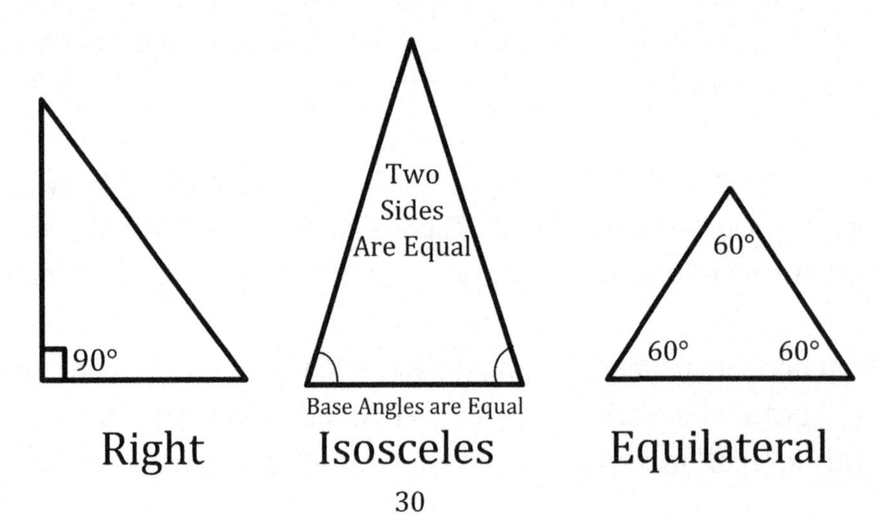

Name: ______________________ Date: ______________

WORKSHEET 4-4

1. Name each type of special triangle pictured below.

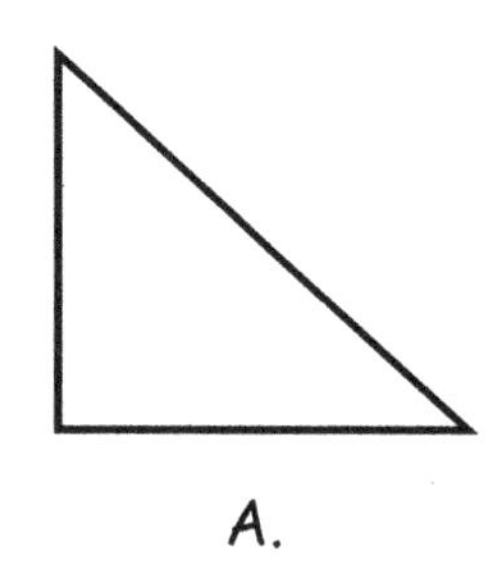

A.

B.

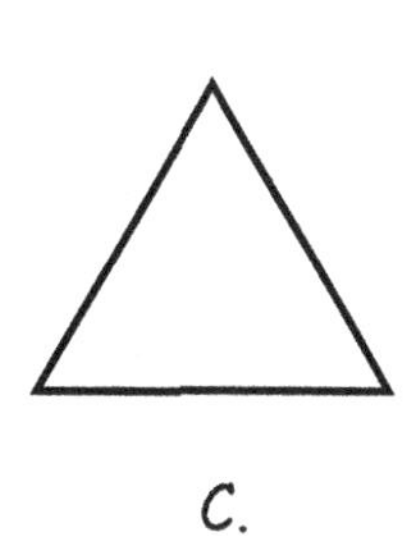

C.

2. Below is an isosceles triangle. Answer the following questions.

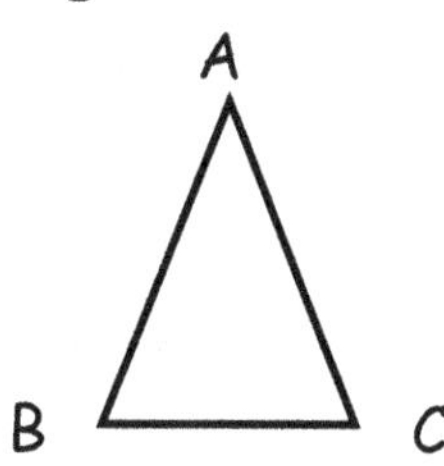

A. Angle B is 70°. What is the measurement of angle A?
B. Side AC is 7". What is the length of side AB?
C. Angle A is 80°. What is the measurement of angle C?

3. Below is an equilateral triangle. Answer the following questions.

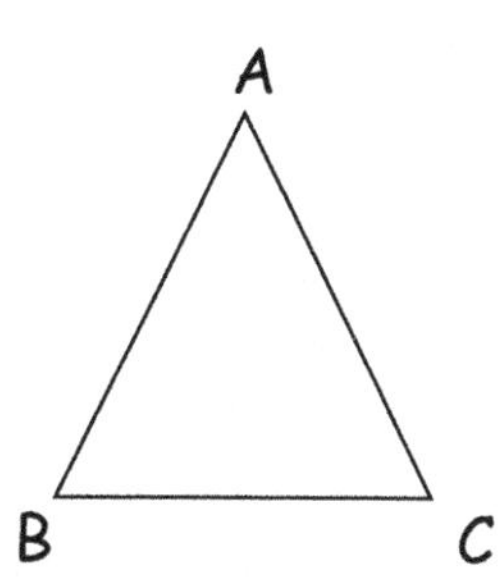

A. What is the measurement of angle B?
B. Side AC is 10'. How long is side BC?
C. What is the total measurement of angles A, B, and C together?

WORKSHEET 4-4 page 2

4. Below is a right triangle. Answer the following questions.

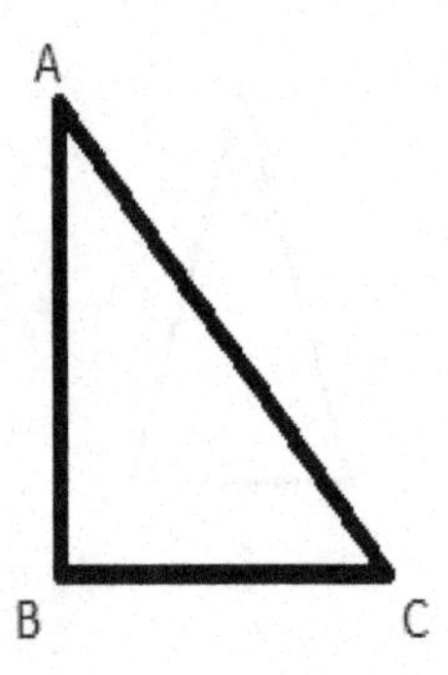

A. What is the measurement of angle B?
B. If angle C is 45°, what is the measurement of angle A?
C. If angle A is 50°, what is the measurement of angle C?

5. All triangles have 3 angles. If you measure all 3 angles and add them together, what will they total?

6. I'm thinking of a triangle. Two of the angles are 60°. What kind of triangle is it?

7. I have a right triangle. One of the angles measures 50°. What are the measurements of the other two angles?

8. I have a triangle. The measurements of the angles are 30°, 60°, and 90°. What type of triangle do I have?

9. What does it take for a triangle to be called an equilateral triangle>

10. What does it take for a triangle to be called an isosceles triangle?

11. Is it possible for one triangle to be all three special triangles?

LESSON 5: PLANE FIGURES

Next we will learn about *plane figures.* Flat shapes are called *plane figures.* The square and rectangle below are flat plane figures. A piece of paper is a plane figure because it is flat; it's a 2-dimensional shape. A 2-dimensional shape has only 2 dimensions; the height and the base. A *dimension* is a measurement. A box is NOT a plane figure. It is a 3-dimensional shape. It has 3 dimensions; height, base, and width. That's what 3D means, three dimensions or three measurements.

Look at the two plane figures below. They are a square and a rectangle.

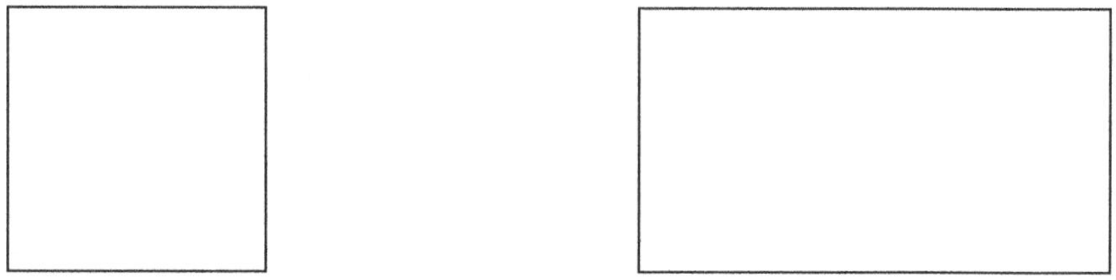

Square

Rectangle

They both have four sides and four 90° angles. But, in order to be called a square, all four sides must be the same length. A rectangle can have two sides longer than the other two.

In order for a 4-sided shape to be a square or a rectangle, the sides must be *parallel* to each other. The two lines below are *parallel* to each other.

These two lines are spaced apart evenly. Even if we drew them super long, they would never cross each other because they are parallel. The two lines below are not parallel.

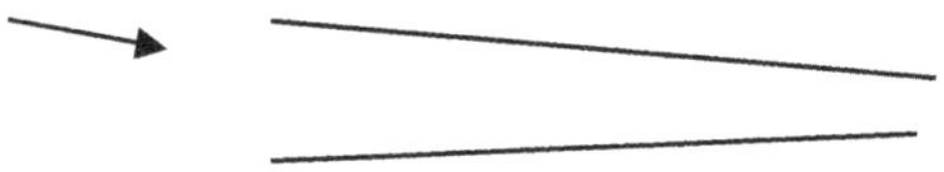

If we drew those lines longer, they would eventually cross each other, so they are not parallel.

The next word you need to learn is *Perpendicular*. It is completely opposite of parallel.

These two lines are perpendicular to each other.

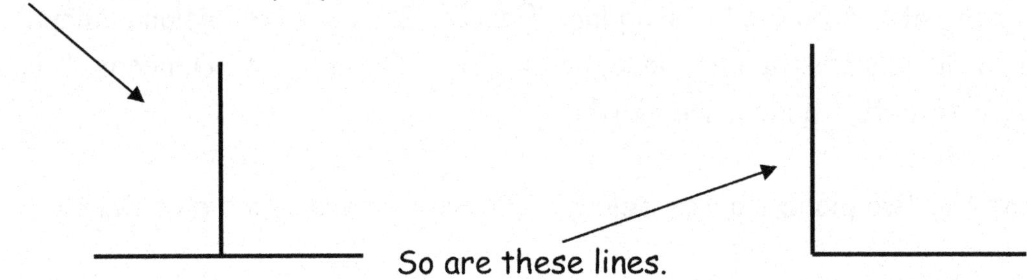

Whenever two lines form a "T" or an "L" shape with a perfect 90° angle, then they are perpendicular to each other.

If you can correctly answer the questions on the following worksheet, then move on to the next lesson. If you get any wrong answers, read the last few pages again.

Name:______________________________ Date:______________

WORKSHEET 4-5

1. Think of all the numbers between 1 and 20. Which number looks like two parallel lines?

2. Which two letters of the alphabet look like perpendicular lines?

3. I have a shape with four 90° angles. Two of the four sides measure 2", the other two sides measure 4". What kind of shape do I have?

4. Which set of lines are parallel to each other?

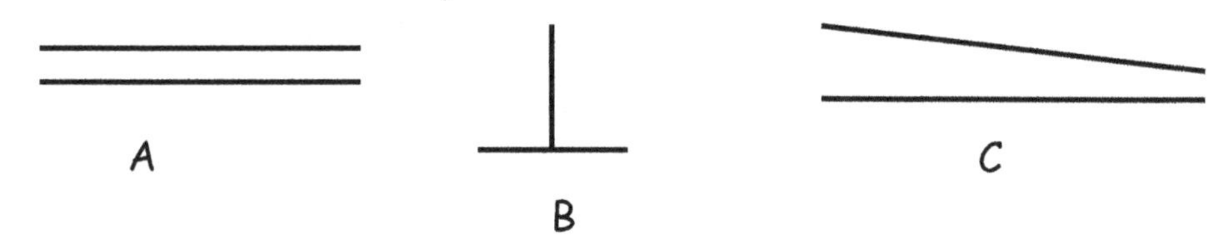

5. Look at A, B, and C above in problem number 4. Which set of lines are perpendicular to each other?

6. Which angle is a 90 degree angle?

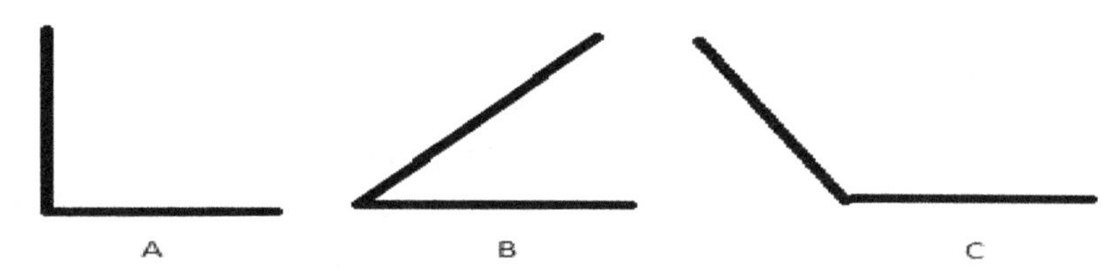

7. Which shape is a square?

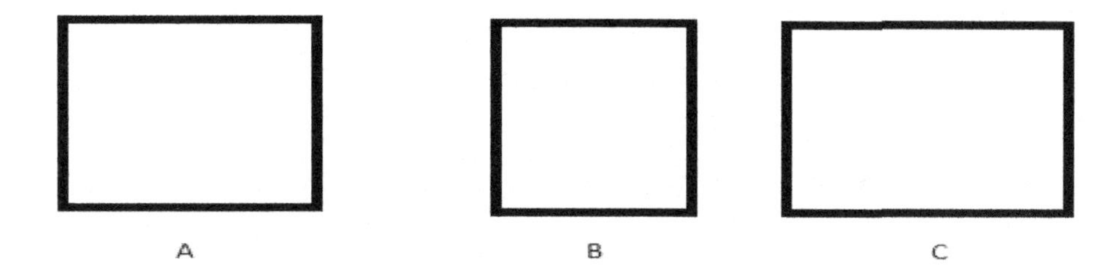

8. How many dimensions does a plane figure have?

LESSON 6: AREA OF A SQUARE

The next new word you need to learn is *area*. *Area* is best described as an area rug. When you are asked to find the *area* of a shape, they want to know what size rug it would take to cover the shape. For example, this rectangle is 3 inches tall by 5 inches wide.

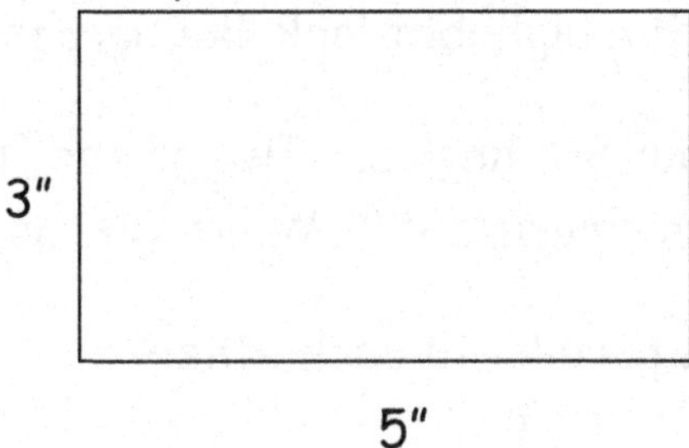

3″

5″

To find out what size rug you would need to cover this square, you just multiply the two numbers.

5″ x 3″ = 15 square inches

Since you are multiplying "inches x inches" the unit in the answer is "square inches." You can write the answer as 15 square inches or $15in^2$, either way is correct.

Look in your Geometry Kit.

> If you don't have the Geometry Kit you can purchase one on our website or just cut out 15 little squares that measure 1 inch on all 4 sides. Then get a 3 x 5 index card and cover it up with the 15 little squares.

Find the 1″ x 1″ foam squares in the Geometry Kit. Each square is 1 square inch. Look for the 3″ x 5″ card in your Geometry kit too. You can fit 15 of those little 1″ by 1″ squares onto that 5″ x 3″ rectangle. Put your 1″ squares over the 3″ by 5″ rectangle. It takes 15 of the little squares to cover the entire card. This proves that the 3″ x 5″ card has 15 square inches of area.

If the card were only 2″ x 5″, then it would only take 10 of those squares to cover the shape. The area of a 2″ x 5″ rectangle is 10 square inches.

This next rectangle is 6 feet tall by 8 feet wide (not really, of course). When you multiply these two numbers, the answer will be in square feet.

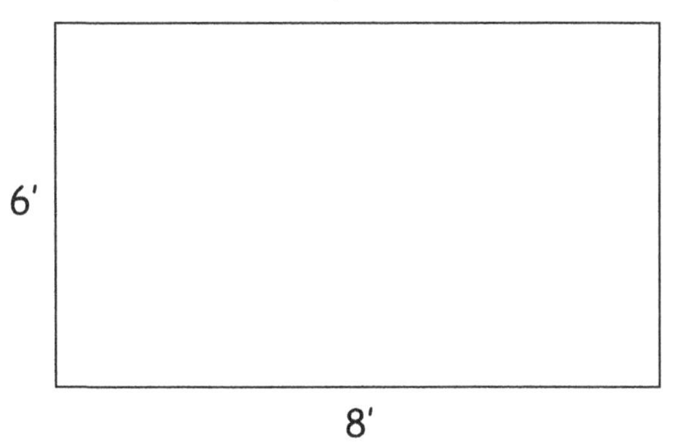

To find the area, multiply the height x the width. This formula is written as Area = base x height. Those are the two dimensions of our 2-dimensional plane figure. The base is 8' and the height is 6'. This is how we write that formula for short.

$$A = bh$$

Let's do the math, 6' x 8' = 48 square feet. The area of the rectangle is 48 square feet. This can be written 3 different ways:

48 square feet or 48 sq.ft or $48ft^2$

You have one square foot piece of paper in your Geometry Kit. If the shape on the last page were actual size, it would be the same size as 48 of those paper squares. If you don't have the kit, cut out a piece of paper, maybe use a newspaper or something, and make it measure 12" on all four sides. That piece of paper is equal to one square foot.

Practice finding the area of the following shapes. Remember to use the Area formula. It is the first formula listed on your Formula Card from the Geometry Kit.

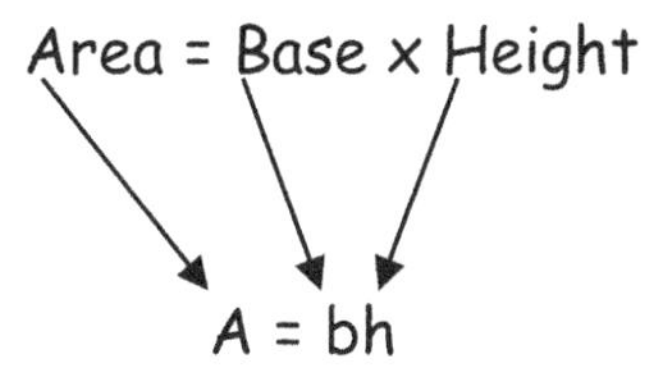

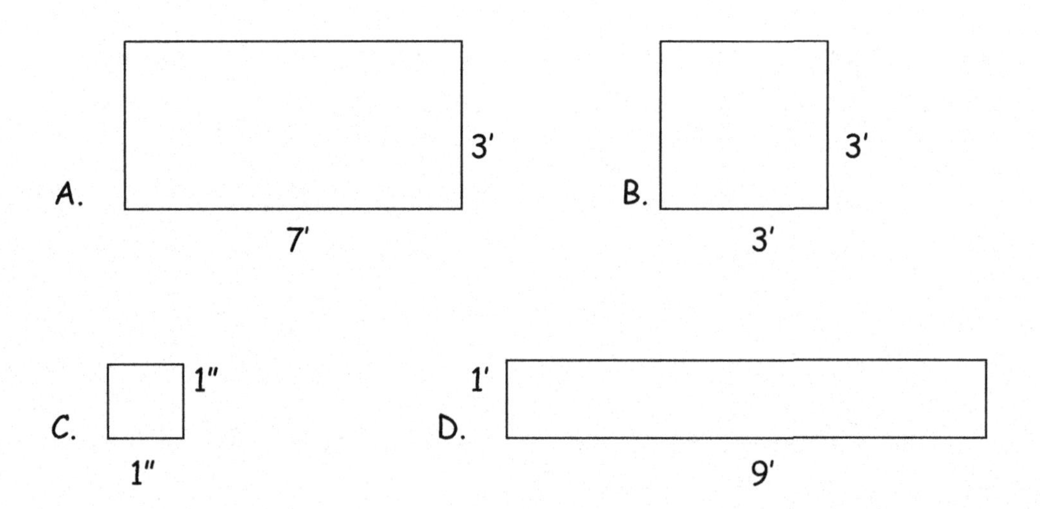

Solutions:

A. Just multiply the base times the height, 7' x 3' = 21 square feet.
B. Multiply the base times the height, 3' x 3' = 9ft^2.
C. This one is easy. You have 15 of these in your math kit. 1" x 1" = 1in^2.
D. Quick, solve this one in your head. The answer is 9 square feet.

Name:______________________________ Date:______________

WORKSHEET 4-6

1. Use A = bh to find the area of this square.

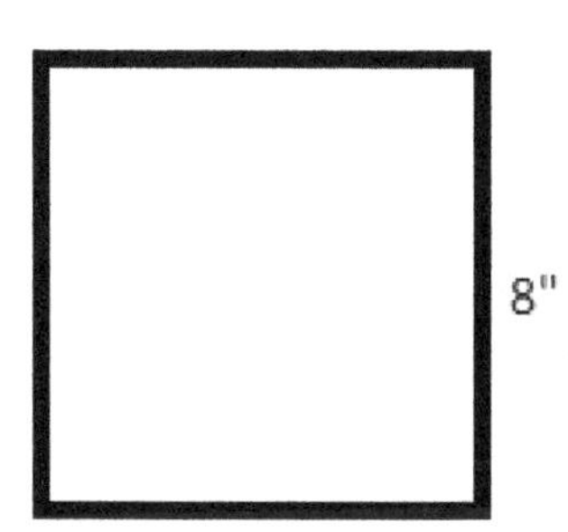

2. Find the area of this rectangle.

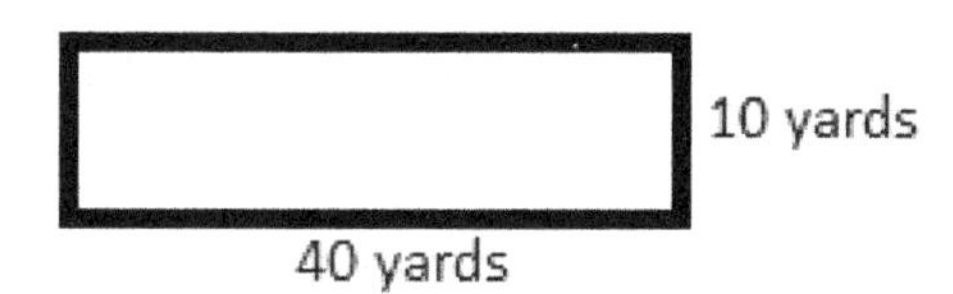

3. Your back yard measures 30 feet x 40 feet. What is the area of the back yard?

4. What is the area of a 9 foot tall square?

5. My Geometry book measures 11" x 8". How many square inches are on the cover?

6. I'm going to put tile on a floor. Each tile is 1 square foot. The floor measures 20' on one side and 24' on the other side. How many tiles will I need to cover the floor?

Next, we will find the area of a triangle. Look in your Geometry Kit and find the square that has been folded in half to form two triangles. It looks like this:

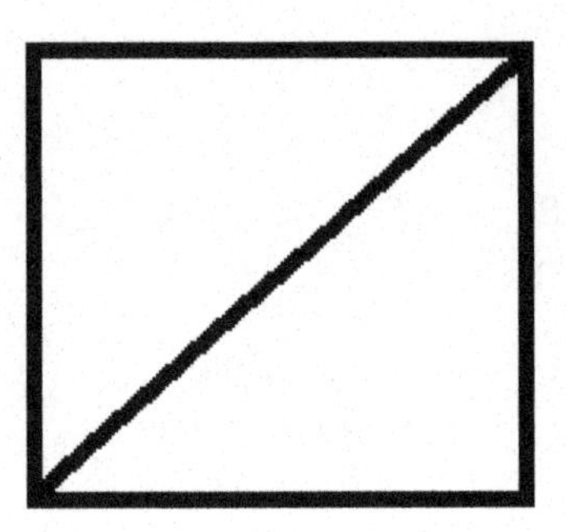

Do you see how a one half of the square is a triangle? So, when it comes to finding the area of a triangle, we still multiply base x height, but then we cut that number in half because a triangle is only half of a square (or rectangle).

I realize the triangle in that last example is a right triangle, but the theory is true for all triangles. The next picture is an isosceles triangle inside a rectangle. They both have a base of 4". If you were to cut off the gray triangles, they would fit perfectly over top of the white triangle. That proves that the area of a triangle is always half the area of the square or rectangle that it fits into.

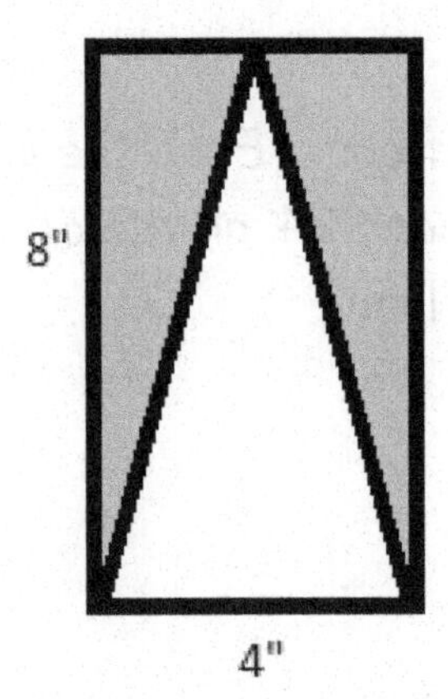

The formula for finding the area of a triangle is $A = \frac{1}{2}bh$. It is the second formula on your Formula Card. Look at the next triangle. We will find the area together.

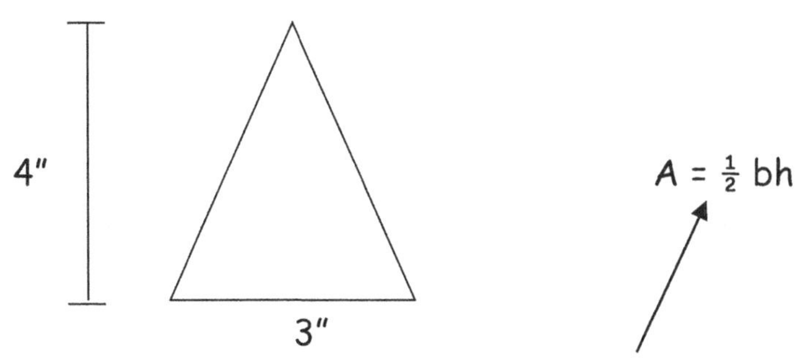

To find the area of a triangle use this formula. For this triangle, base times height would be $3 \times 4 = 12$. If this were a whole square, the area would be 12 square inches. But a triangle is only half a square, so we need to cut that number in half. $12 \times \frac{1}{2} = 6$. The area of the triangle is 6 square inches.

Try again, this time we will find the area of a right triangle. Use the same formula that we used last time, $A = \frac{1}{2}bh$.

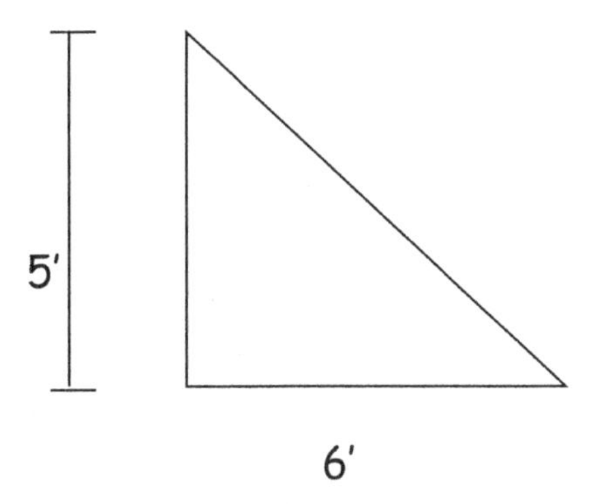

To find the area of any triangle, multiply the base times the height and then cut that number in half. That's what the formula is saying, "Area = one half base times height." For the right triangle above, multiply $5 \times 6 = 30$. Cut that number in half and the area of the triangle is 15 square feet. Next we will find the area of this equilateral triangle.

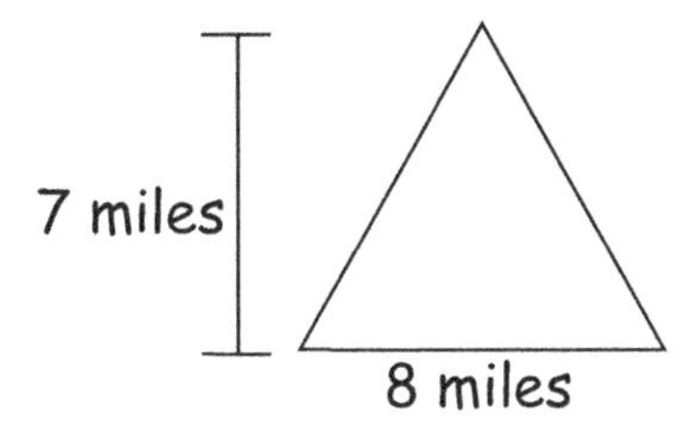

To find the area of any triangle use the formula $A = \frac{1}{2}bh$. Be sure you are using the "height" and not the "side" of the triangle, those can be two different dimensions. The triangle on the last page is 8 miles on the base and it has a height of 7 miles. Multiply those two amounts together, 8 miles x 7 miles = 56 square miles. Now cut that answer in half. What is the area of this triangle? You should have answered 28 square miles.

Practice finding the area of a triangle by completing the next worksheet.

Name: ______________________________ Date: ______________

WORKSHEET 4-7

Find the area of the following triangles using the formula $A = \frac{1}{2}bh$.

1.

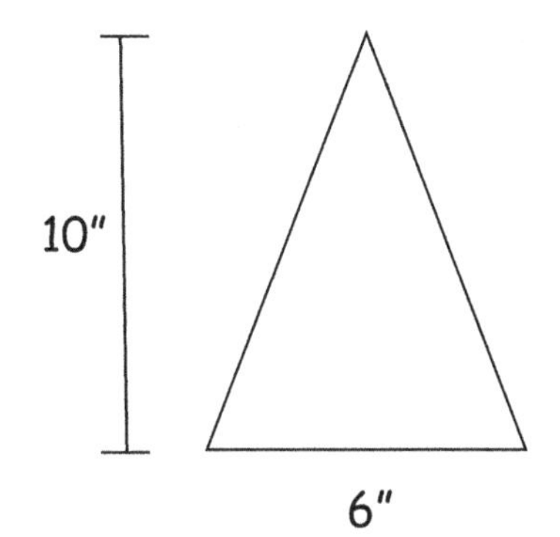

2.

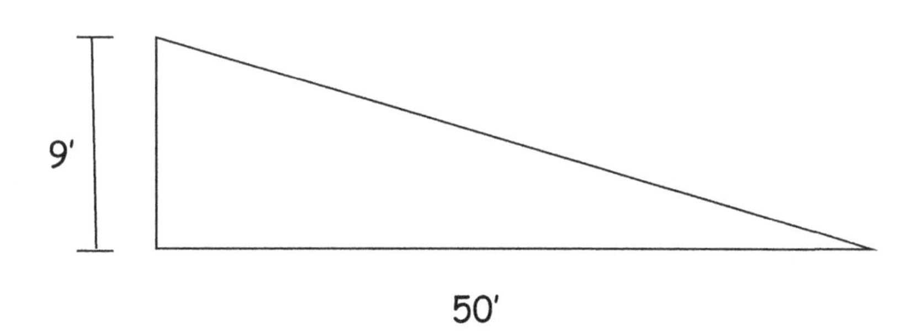

3.

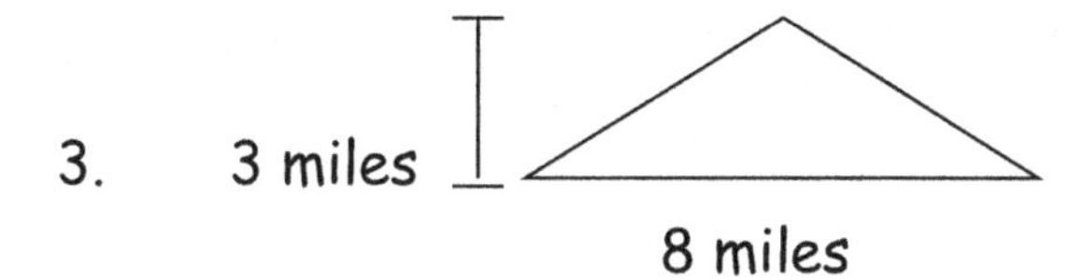

4. Look at triangle ABC and then answer the following questions. You won't need a protractor or a ruler. All you need to know is that it is an **equilateral** triangle and side AC = 4".

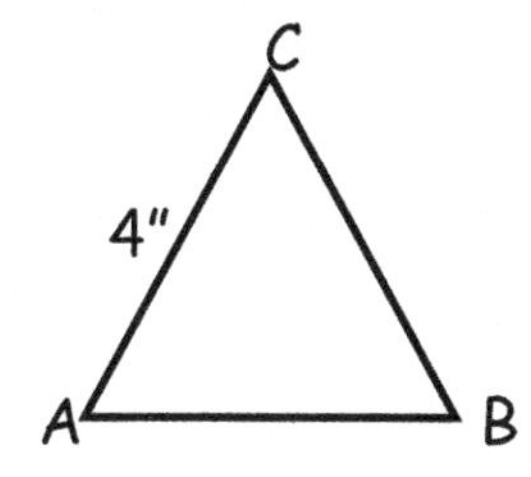

a. What is the length of side BC?

b. What is the size of angle A?

c. What is the length of side AB?

d. What is the size of angle C?

5. Look at triangle DEF and answer the following questions. You won't need a protractor or a ruler. All you need to know is that it is a **right** triangle.

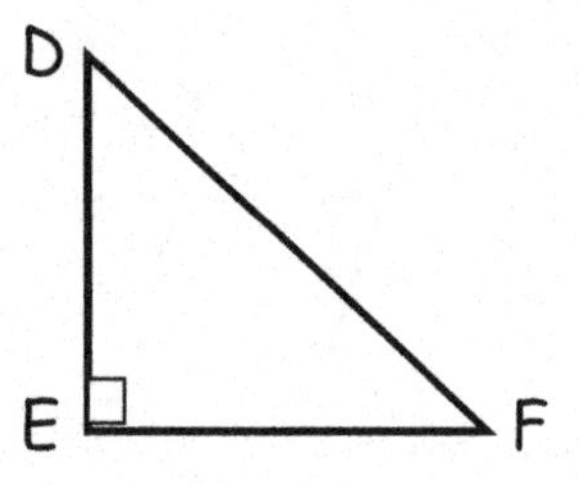

a. What size is angle E?

b. Angle D is 45°. What size is angle F?

c. What does the little box in angle E mean?

6. I'm going to paint a triangular shape on the wall. The triangle will be 8 feet tall. The base of the triangle will be 5 feet. How many square feet of wall will I be painting?

7. If I fold any square in half diagonally, what two new shapes will I get?

Check your answers and see how you did. If you got any of the answers wrong, go back and find what you missed. Once you understand everything on this worksheet, you are ready to move on.

In those last few lessons, you learned that a triangle is a 3-sided shape. You also learned that squares and rectangles are 4-sided shapes. In this lesson, you will learn about *polygons*. A *polygon* is a shape with any number of sides. Of course, the sides have to be straight and connect at each end, but it can have any number of sides, well three or more. You can't draw a shape with only two sides; that would be an angle.

What is the name of a 3-sided polygon? It's a triangle. What is a name for a 4-sided polygon? A square is one type of 4-sided polygon, but ALL 4-sided polygons are called *quadrilaterals;* quad means 4.

Do you know the name for a 5-sided polygon?

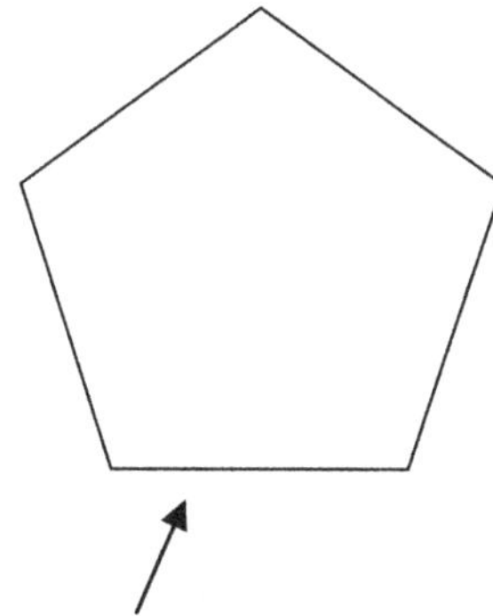

A 5-sided polygon is called a *Pentagon.* There is a Government building in the state of Virginia called The Pentagon. It is built in the same shape as the pentagon above. You should look it up sometime; it's a very cool looking building. That's how I remember what a pentagon looks like; The Pentagon.

Have you ever heard of a Hexagon? A hexagon is a 6-sided polygon. I'm sure you've seen one before; it is a very common shape.
This one is easy to remember because it is the only shape name with the letter "x" in it, just like the number 6 is the only number with the letter "x."

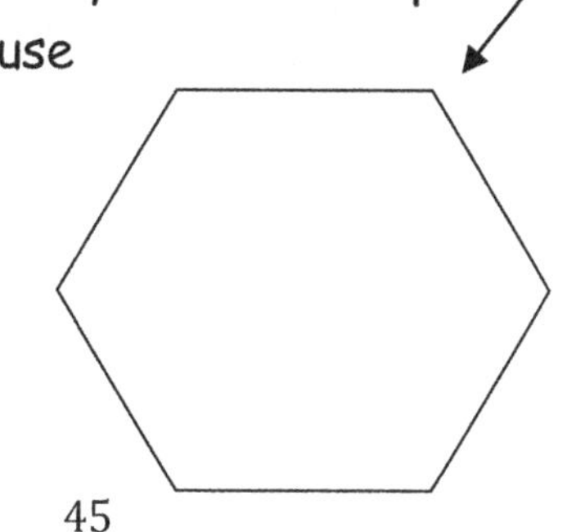

A 7-sided polygon is called a *heptagon*. But it is so uncommon, don't even bother trying to memorize that one. However, an 8-sided polygon is very common, it is called an *octagon*. A stop sign is an octagon shape. It's easy to remember that an octagon has 8 sides because an octopus has 8 legs; "oct" means 8.

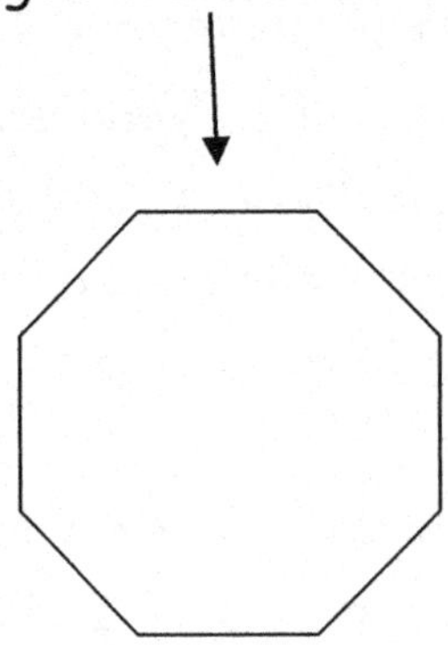

There are names for 9-sided polygons and 10-sided polygons, but these two are not very common either. There is a lot to remember in math, but knowing that a 9-sided polygon is called a *nonagon* or that a 10-sided polygon is called a *decagon*, is not very important to remember.

What you do need to remember is that a polygon is an enclosed shape, made of only straight lines that don't cross each other. Each polygon gets a special name depending on how many sides it has. Here are the most common polygons:

A triangle has 3 sides
A quadrilateral has 4 sides
A pentagon has 5 sides
A hexagon has 6 sides
An octagon has 8 sides

Complete the next worksheet.

Name: ________________________________ Date: ______________

WORKSHEET 4-8

Name each polygon.

1.

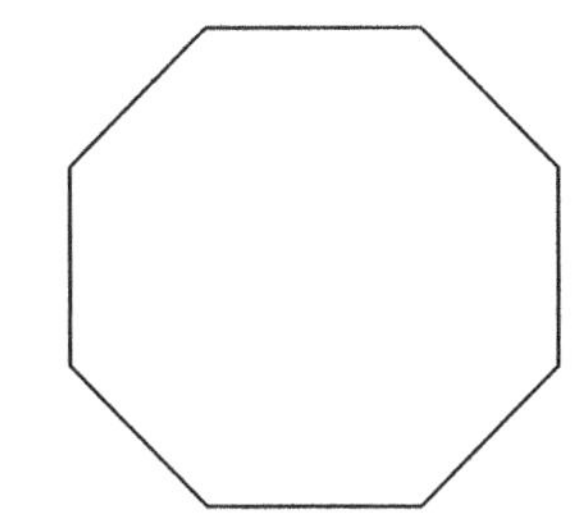

2.

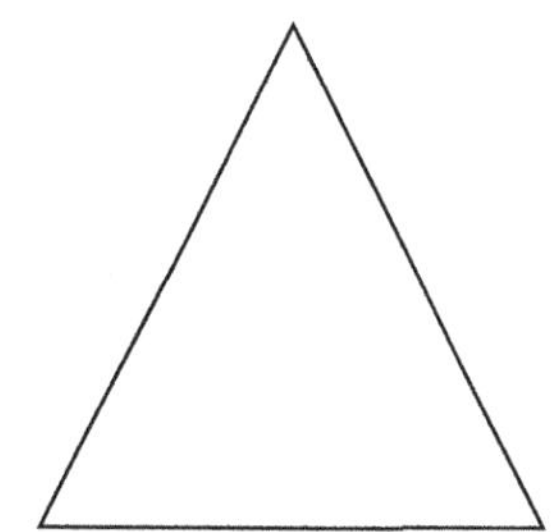

3.

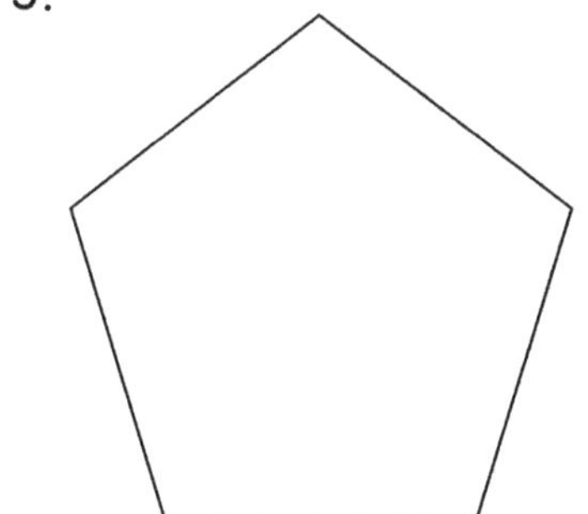

4.

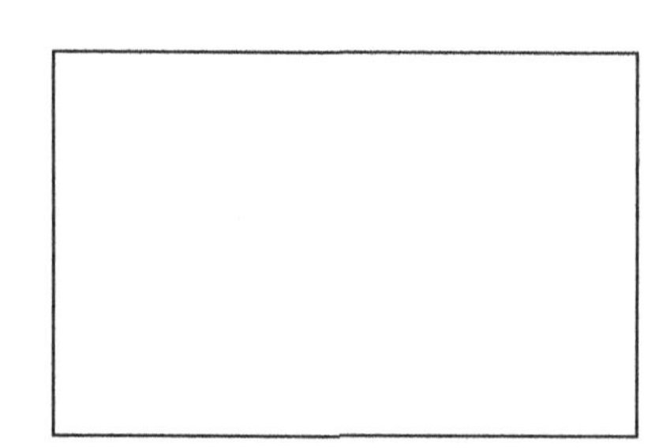

5.

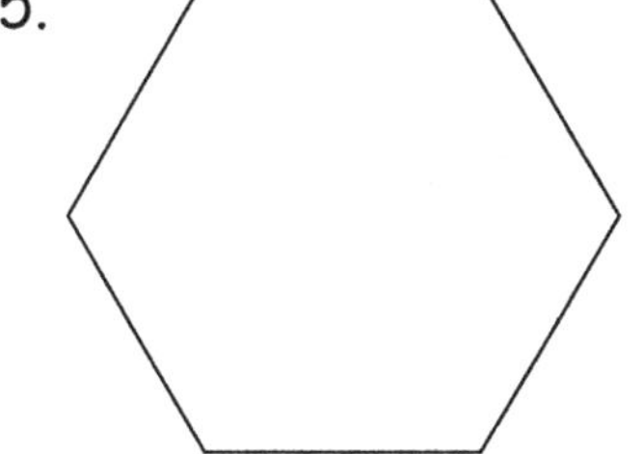

6.

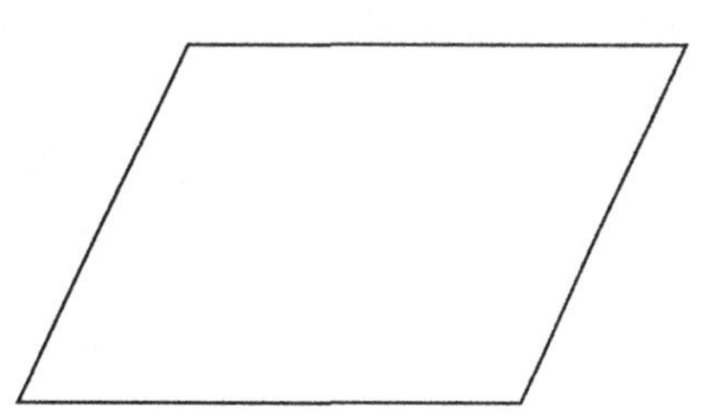

7.

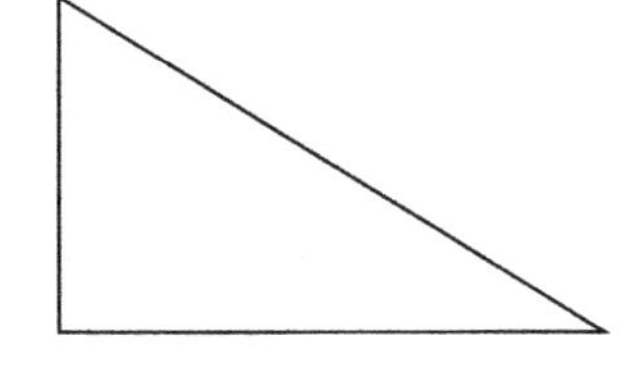

8.

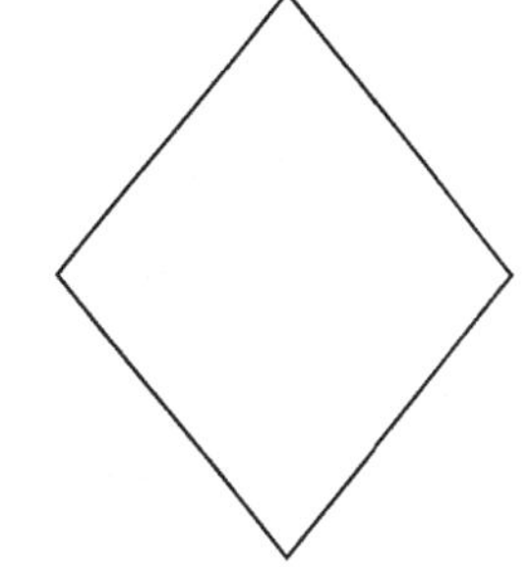

Name: ______________________ Date: ______________

CHAPTER 2 REVIEW TEST

Name the following shapes:

1. 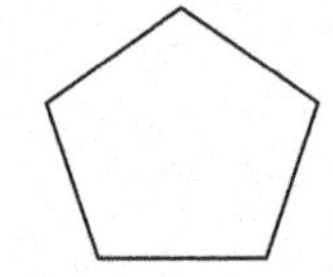

2. 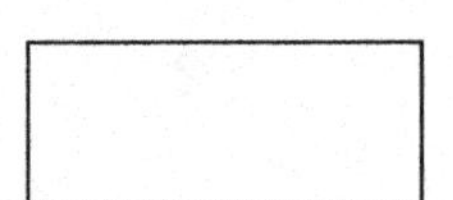

3. 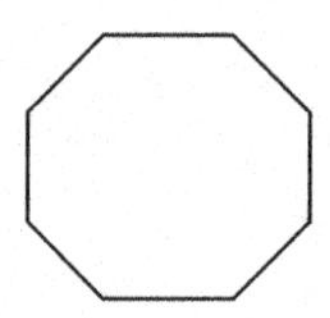

Find the area of the following shapes:

4. 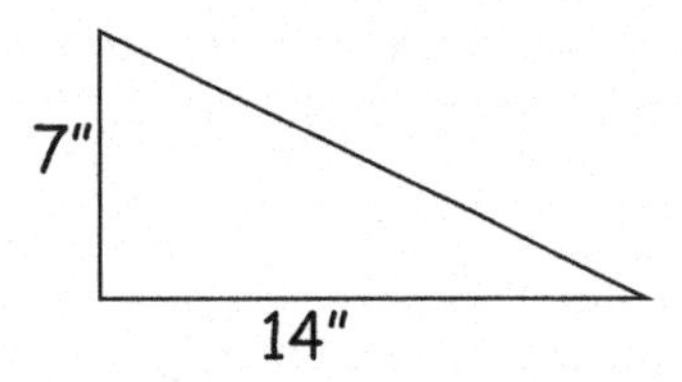

5. 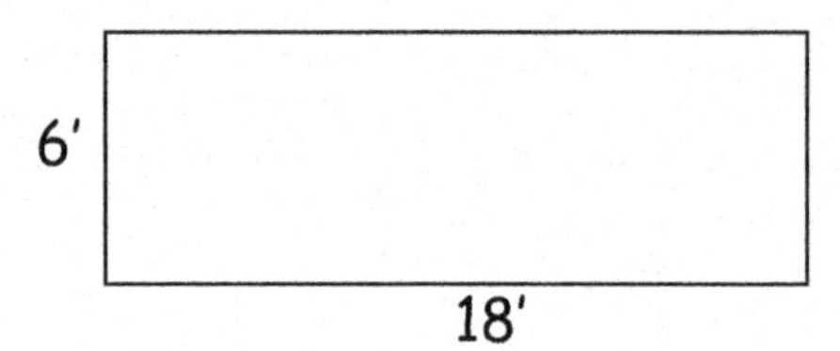

6. 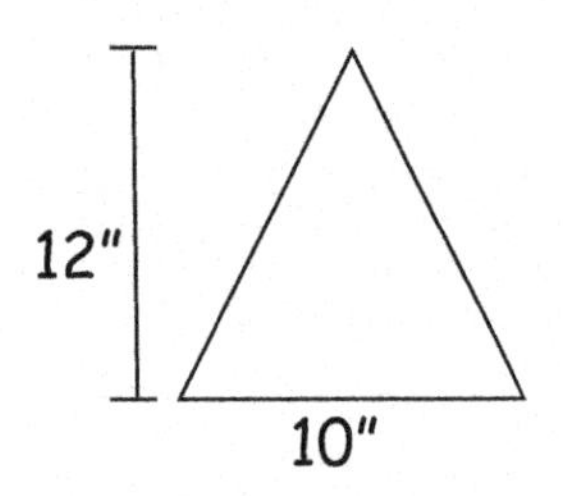

7. 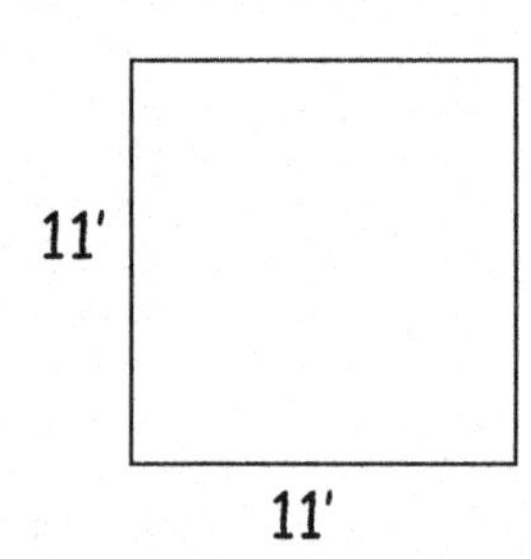

8. Name the 3 special triangles.

9. Which of the 3 special triangles has 60° angles on all 3 sides?

10. Which of the 3 special triangles has at least 2 equal sides?

11. Which of the 3 special triangles has a 90° angle.

12. What is the total of all three angles in an isosceles triangle?

13. Brianna drew a plane figure on a piece of paper. The figure has 4 angles with opposite sides being parallel. The top and bottom are both 6″ long. The 2 sides are both 8″ long. What type of shape did Brianna draw?

CHAPTER 3

PYTHAGORAS

LESSON 9: HYPOTENUSE

Let me tell you a few clues about right triangles that only smart people know. But first, you must learn one word. Let me explain. Earlier you saw a right *angle* get turned into a right *triangle* with just one line.

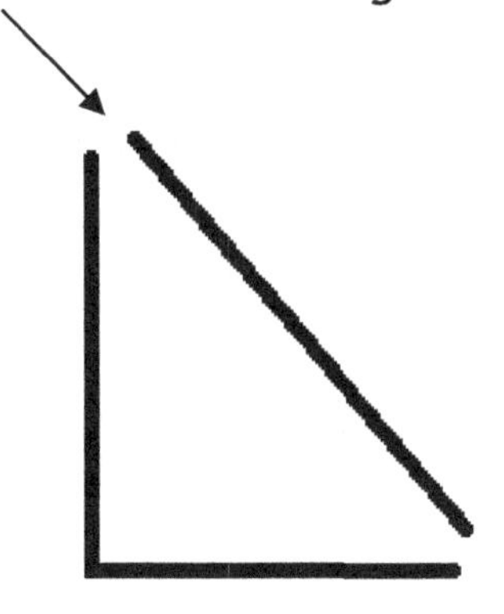

This line is called the *hypotenuse* (high-pot-a-noose) of the right triangle. Look at the right triangle below.

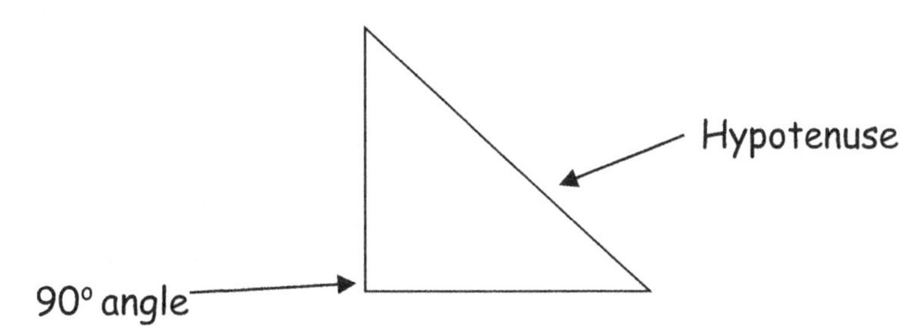

The hypotenuse is the side of a right triangle that is not part of the 90° angle. Some people like to look at the hypotenuse as the longest side of the right triangle. Find the hypotenuse of each of these right triangles.

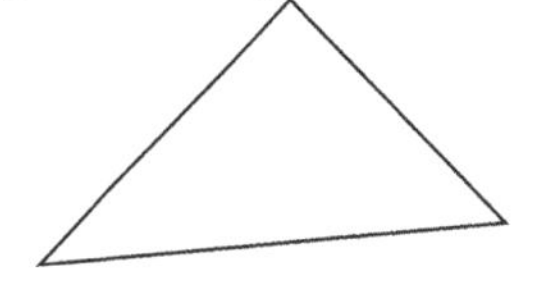

To find the hypotenuse, first find the 90° angle. Next, find the side that isn't part of that angle or the third line that turns the 90° angle into a triangle.

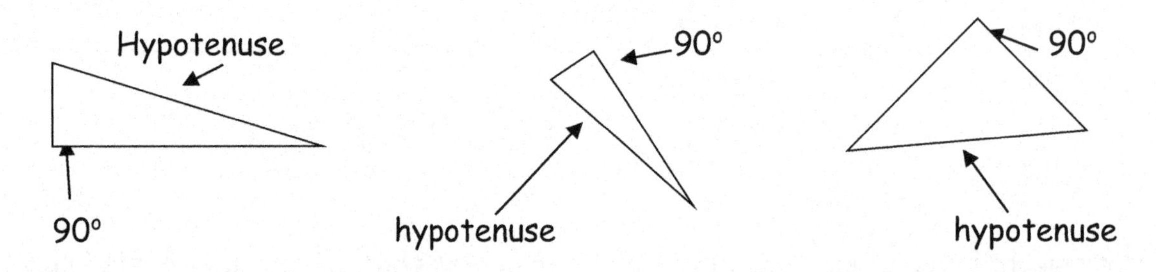

If you understand which side of the right triangle is the hypotenuse, I will explain the big math secret that only smart people know.

A long, long time ago a guy named Pythagoras figured out a way to find the length of a hypotenuse. As long as he knew how long the base was and how tall the 90° angle was, he could figure out the length of the hypotenuse.

Pythagoras figured out that if you square this number, (3 x 3 = 9).

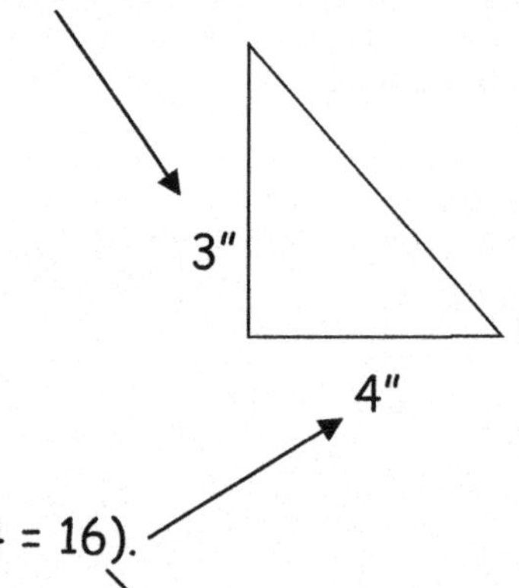

And you square this number (4 x 4 = 16).

Then add those two answers together (9 + 16 = 25). You will be able to figure out the length of the hypotenuse.

All you have to do is find the square root of 25 and you will have the length of the missing side. This formula is called the "Pythagorean Theorem" (Pi-thag-or-en). This formula is only for right triangles.

I know it sounds confusing, but it's not as complicated as it sounds. Just keep reading.

All right triangles have three sides. We will call them sides a, b and c. The hypotenuse will always, always, ALWAYS be labeled side C. Look at the right triangle below. The hypotenuse is side c.

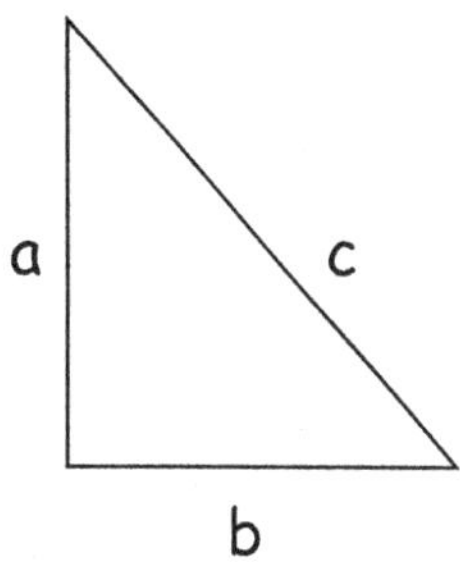

The Pythagorean Theorem is: $a^2 + b^2 = c^2$

So, let's try it out and see if Pythagoras was right. If you are unclear on squaring numbers or finding the square root of a number, read Volume III of the *Learn Math Fast System.*

As a reminder, to square a number is to multiply it by itself. For example, 3^2 means 3 x 3, so $3^2 = 9$. Finding the square root of a number is the reverse math. For example, 3 x 3 = 9 and the square root of 9 is 3. Another example is, 8 x 8 = 64. Now reverse that - the square root of 64 is 8.

Here is a right triangle.

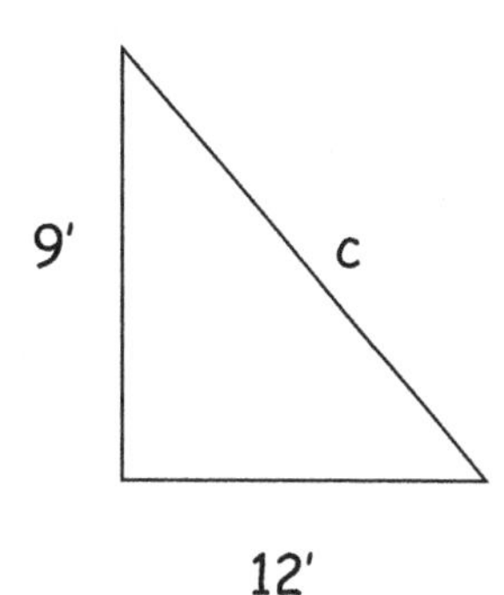

Plug those numbers into our formula and do a little algebra. If you don't understand algebra, read Volume III of the *Learn Math Fast System*.

$$a^2 + b^2 = c^2$$

$$9^2 + 12^2 = c^2$$

$$81 + 144 = c^2$$

$$225 = c^2$$

To get rid of the ^{2}, find the square root of c^2.
Whatever you do to one side, you must do to the other side.

$$\sqrt{225} = \sqrt{c^2}$$

$$15 = c$$

We just figured out that the hypotenuse is 15 feet long. Don't over complicate the Pythagorean Theorem. It is just a cool thing that Pythagoras figured out a long, long time ago. He says that you can figure out how long the hypotenuse is by squaring the other two sides, adding them together, and then finding the square root of that number.

Try it out yourself on the next worksheet.

Name:______________________________________ Date:__________________

WORKSHEET 4-9

1. Use the Pythagorean Theorem to find the length of this hypotenuse.

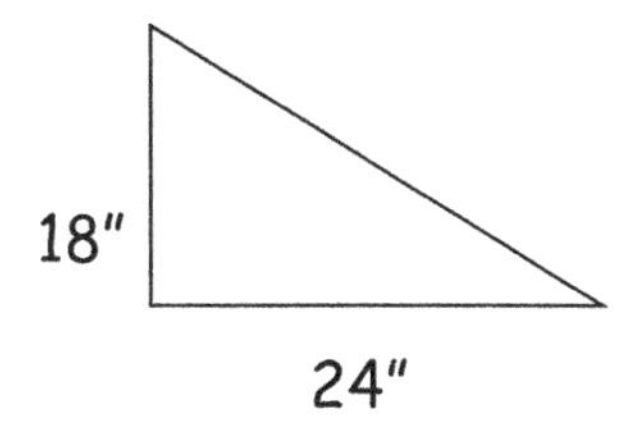

2. What type of triangle has a hypotenuse?

3. Triangle ABC is an isosceles triangle. What is the length of side AB?

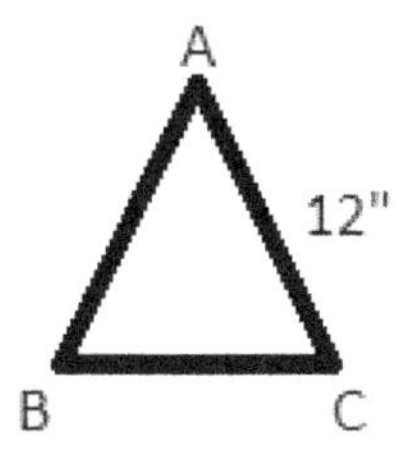

4. Find the length of the hypotenuse for triangles A, B, and C.

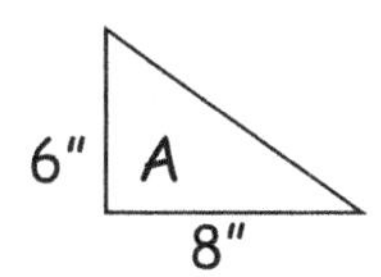

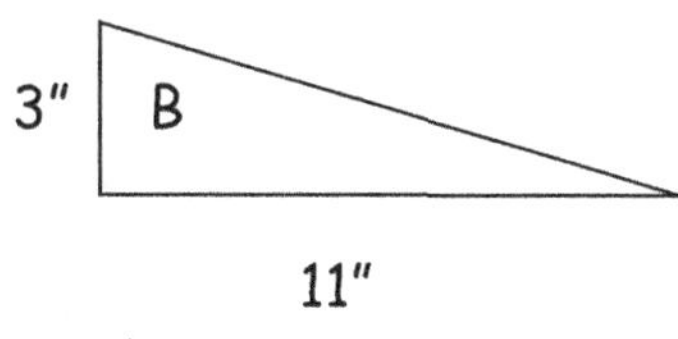

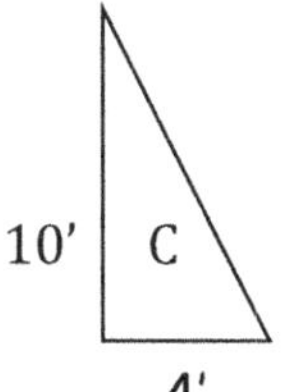

LESSON 10: 3-4-5 TRIANGLE SECRET

To further explain the Pythagorean Theorem, let me show you an easy short-cut way. It's called the 3-4-5 triangle secret.

Look at this right triangle. The sides measure 3, 4, and 5.

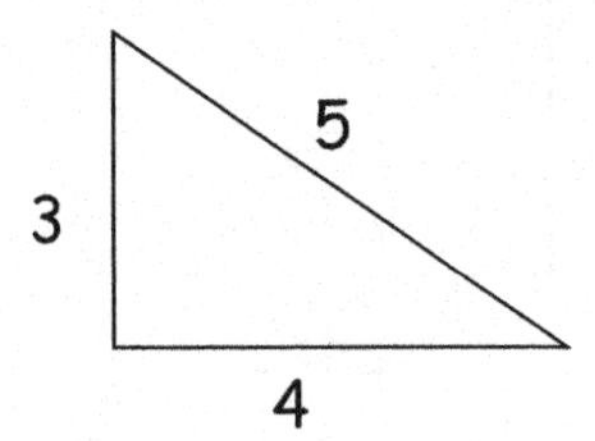

I will plug those numbers into the Pythagorean Theorem.

$$a^2 + b^2 = c^2$$
$$3^2 + 4^2 = 5^2$$
$$9 + 16 = 25$$

Do you see how the math works out perfectly? If I wrote that the hypotenuse measured 6, then the math in the Pythagorean Theorem would not work out, so those dimensions will never form a perfect right triangle. The 3-4-5 triangle is considered a perfect right triangle because the dimensions of the sides make sense when plugged into the Pythagorean Theorem.

Why is that such a big deal? Well, let me show you how that can be helpful. Below is another right triangle. The sides measure 6, 8, and 10.

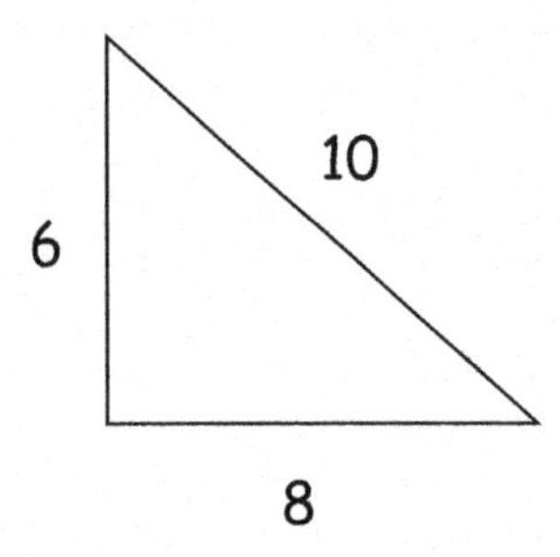

All I did was just double the numbers from the last perfect right triangle, so instead of a 3-4-5 triangle, we have a 6-8-10 triangle. I can tell that this is still a perfect right triangle by putting our numbers into the Pythagorean Theorem. $6^2 + 8^2 = 10^2$ (36 + 64 = 100). (If the hypotenuse had been, let's say 11, the math would not have worked out. It would have been 36 + 64 = 121, which is not true. If the numbers don't fit into the Pythagorean Theorem, then we know the angle is not really 90°, so it is not a perfect right triangle). Now I will double *those* lengths.

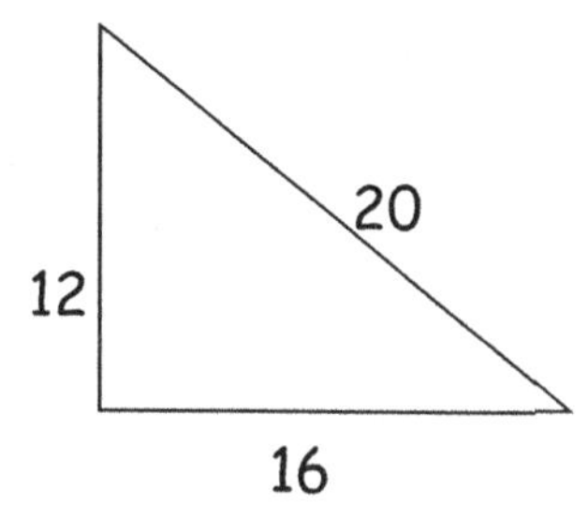

Is this a perfect right triangle? Let's plug the numbers into the Pythagorean Formula. $12^2 + 16^2 = 20^2$ (144 + 256 = 400) Yep, the math pans out, so it is perfect. As long as our lengths are divisible by 3, 4, and 5, with 5 being the hypotenuse, it will always be a perfect right triangle.

This information isn't just for math class. The 3-4-5 triangle secret can save you hours of aggravation the next time you have to make a perfect right angle without a protractor.

Let's say you want to build a fence and you want it to be at a 90° angle from the house. A protractor is too small. It cannot help you. But knowing about a 3-4-5 triangle will.

Let's say the fence you want to build is 30 feet long. And you *don't* want it to look like the drawing on the next page.

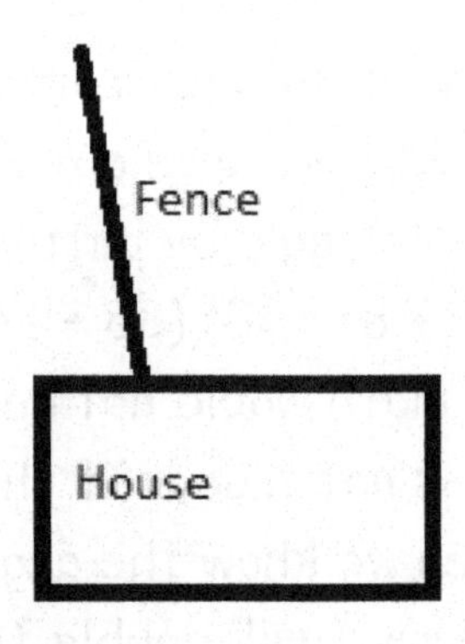

You don't want it to look like this either.

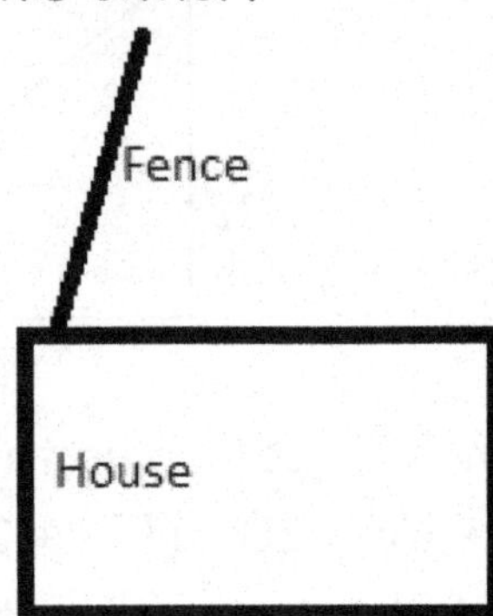

You want the fence to be perpendicular from the house; in a perfectly straight 90° angle, like the fence in the next picture.

To find out where your fence post should go, use the 3-4-5 triangle secret as an invisible protractor.

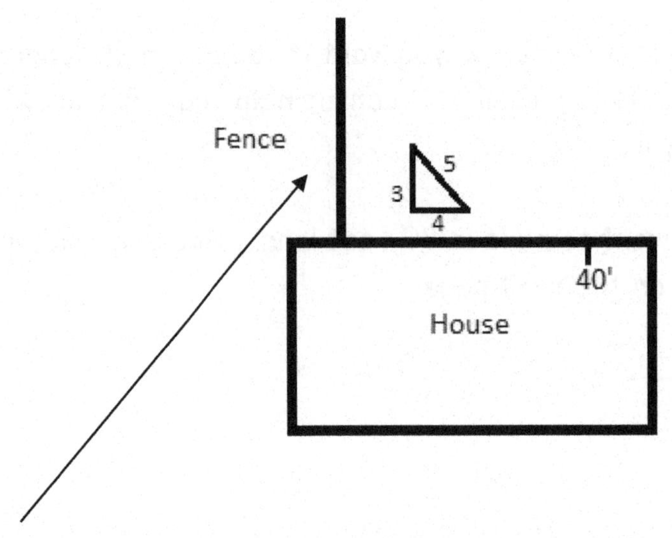

The fence is going to be 30 feet long. That will represent the "3 side" in our triangle. We are going to create a perfect right triangle. We will be using the fence and the house as sides a and b. The hypotenuse, side c, will be an imaginary line whose length we will figure out mathematically.

Once we know the length of the hypotenuse, we can adjust our fence post until it is exactly the right distance away; creating a perfect 90° angle.

The fence (side a) is 30 feet long. That will represent the "3" in our 3-4-5 triangle. It is 10 times bigger than 3, so I will multiply all sides by 10 to keep all sides proportional.

The house (side b) represents the "4" side of our 3-4-5 triangle. We are multiplying all sides by 10, so side b is now 40 feet.

3 x 10 = 30
4 x 10 = 40
5 x 10 = 50

Now our perfect right triangle is a 30-40-50 triangle. The fence line is our 30' measurement, the house is our 40' length, and the hypotenuse will be 50'.

Here is how you use that information. Start by measuring a 30' line, perpendicular to the house. Next, measure 40' from the same point along the house. Now measure the hypotenuse. Adjust your first mark until there is 50' between the 30' line and your 40' line.

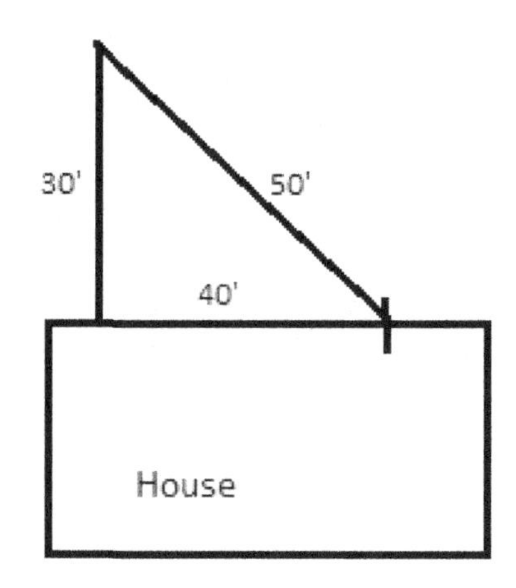

Since the fence is at a perfect right angle from the house, it is considered to be "square" with the house. Do the same thing for the other side of the fence and you will have a perfectly "square" fence.

The Pythagorean Theorem and the 3-4-5 Secret Triangle go hand-in-hand, but deciding which one to use, depends on the project at hand. Let me explain. If you were taking a math test and the problem was to find the length of the hypotenuse below, it would be fastest to use the 3-4-5 method.

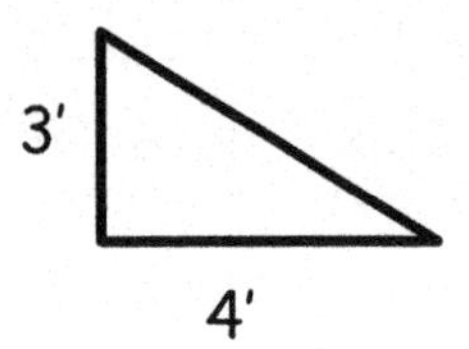

But if the problem were to find the length of the hypotenuse below, then you would want to use the Pythagorean Theorem.

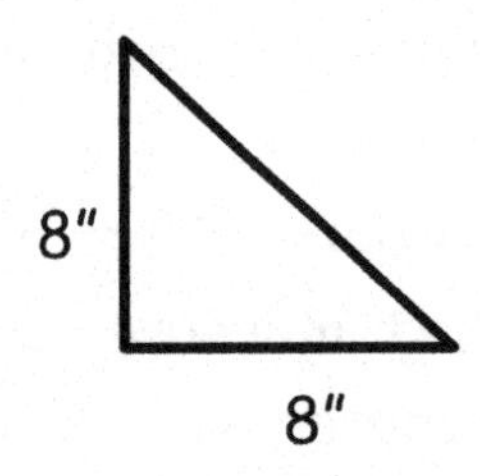

$$8^2 + 8^2 = c^2$$
$$64 + 64 = c^2$$
$$128 = c^2$$
$$\sqrt{128} = \sqrt{c^2}$$
$$\sqrt{128} = c$$
$$11.31" = c$$

The triangle above didn't fit into our 3-4-5 triangle mold, so we had to use the good ol' Pythagorean Theorem.

Now let's talk about real life. Not just a problem on a math test. Let's say you want to rope off a perfect square in the middle of a field. It is next to impossible to do that, unless you know geometry.

It sounds simple, but how can you be sure your square doesn't come out looking like this? Or like this?

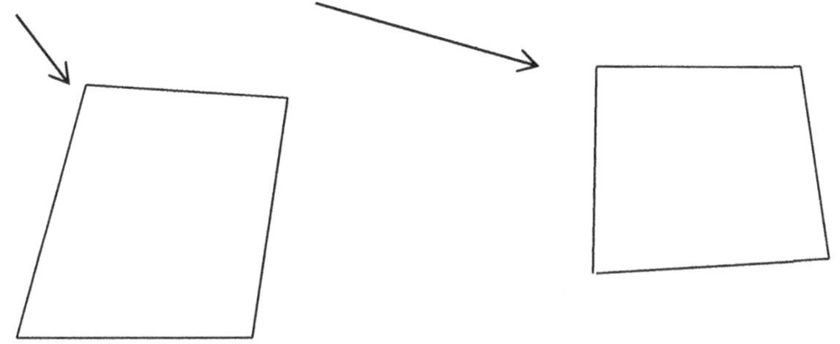

Try to draw a square in the center of a piece of paper without using a protractor or any straight lines.

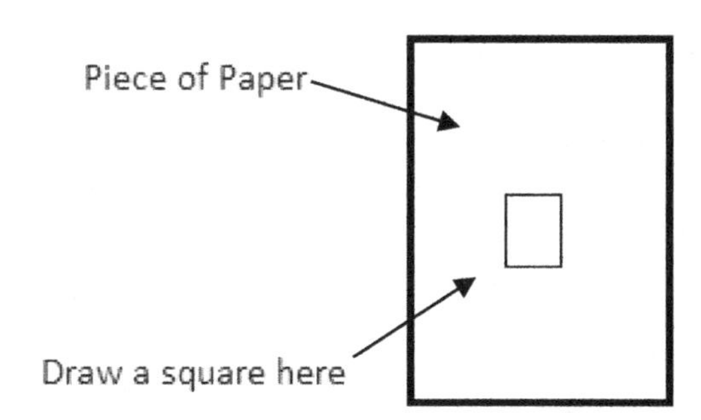

Check your drawing by using a protractor to see if your angles are exactly 90°. It is very difficult to do when you don't have at least one right angle to start with. Now try to draw one using the corner of your paper as guide.

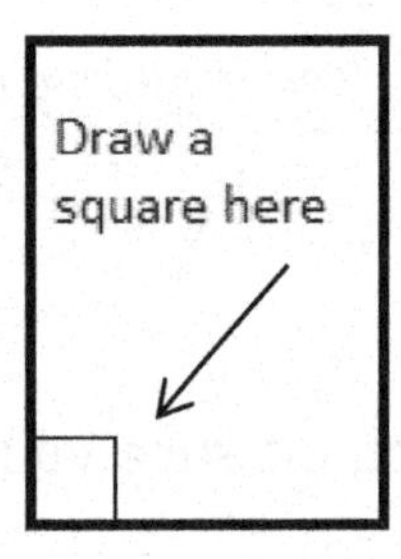

Use your protractor, to find out if your angles are any closer to 90° now. Having at least one perfect 90° angle to work with makes your drawing much more accurate.

The only sure-fire way to draw a perfect 90° angle without any tools is to use the 3-4-5 triangle secret or the Pythagorean Theorem. The only problem with using the Pythagorean Theorem is that sometimes the numbers won't come out even. You can end up with a hypotenuse length of $\sqrt{20{,}000}$ and that won't help you much, when you are in the middle of a field.

Here is what you do. Make your first measurement divisible by 3. For example, if you are trying to make a 100' x 100' square, you could use 99' as the "3" in our triangle. It is very close to 100 and it is divisible by 3.

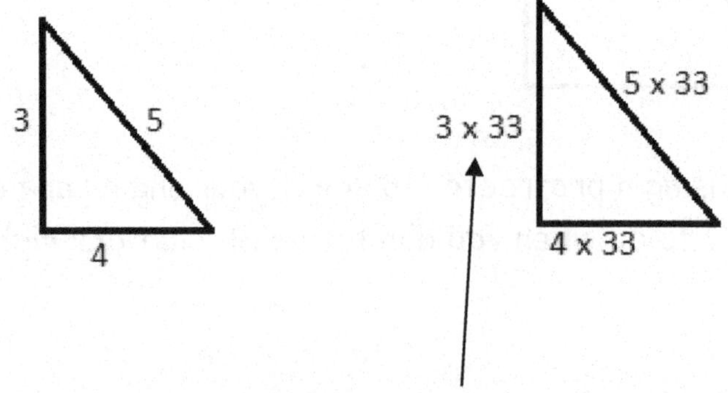

Using 99 means we multiplied our 3 by 33. That is the number we will multiply all sides by. This makes our 3-4-5 triangle a 99-132-165 triangle.

Start by stringing off a 99′ line. Next make a 132′ line as perpendicular as possible.

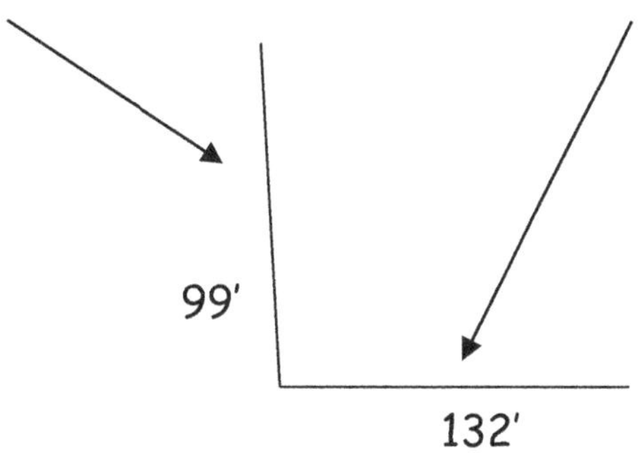

Knowing that our triangle needs to be 99′ x 132′ x 165′ means that you need to make sure there are 165′ between the two ends. Once you adjust the ends of your 99′ and 132′ lines to be exactly 165′ apart, you will have a perfect 90° angle.

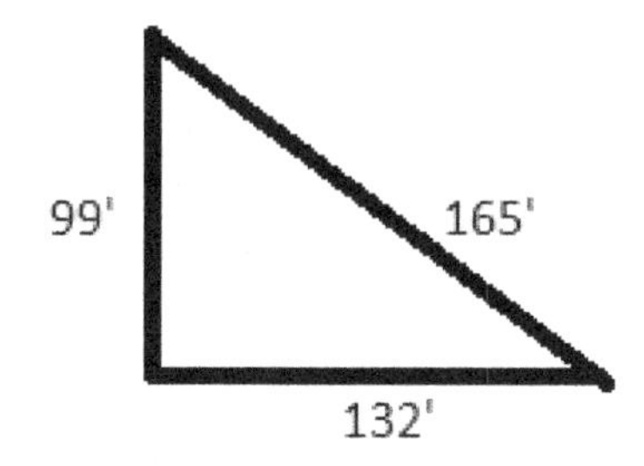

With this information you can easily plot out your 100′ x 100′ square because you have at least one perfect angle to start with.

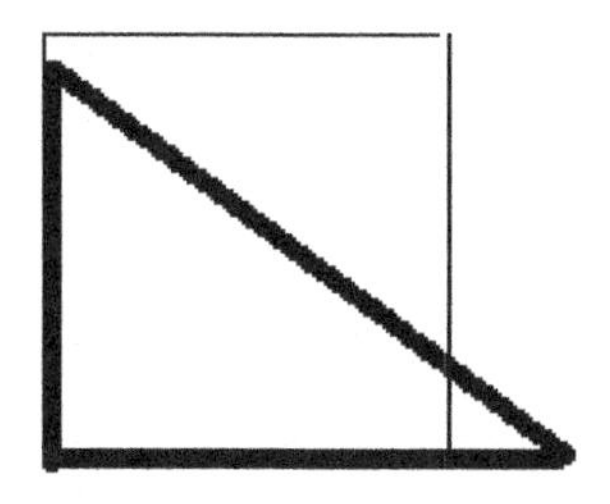

Do you remember when I told you that a triangle is half a square? Below are two triangles, together they create a square, or maybe it's a rectangle, I didn't measure it. Can you find the hypotenuse of the triangles? It's the diagonal line that cuts the square in half.

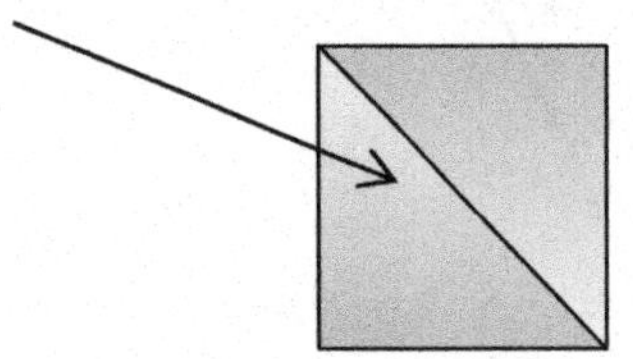

So, finding the hypotenuse of a triangle is the same thing as finding the diagonal of a square (or rectangle).

Try one on your own. Let's say you need to draw a rectangle on the floor. The rectangle needs to be 45" x 60". What is the diagonal measurement for this rectangle?

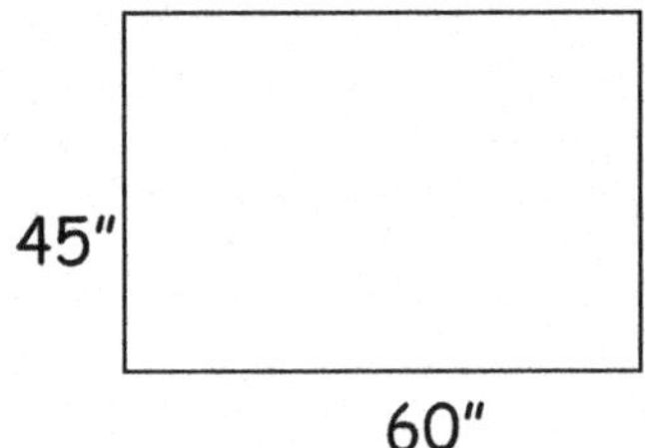

Try to figure it out on your own before reading any further. You can use the 3-4-5 triangle secret or the Pythagorean Theorem to find the diagonal of this rectangle.

The Pythagorean Theorem is $a^2 + b^2 = c^2$. Fill in "a" with 45. Fill in "b" with 60 and do the math.

$$a^2 + b^2 = c^2$$

$$45^2 + 60^2 = c^2$$

$$2025 + 3600 = c^2$$

$$5625 = c^2$$

$$\sqrt{5626} = c$$

$$75 = c$$

Or you could use the 3-4-5 triangle secret. Side "a" or "3" is 45 inches. This number is divisible by 3, so we start there.

$$45 \div 3 = 15$$

That means each side of our 3-4-5 triangle is multiplied by 15. What is the measurement of the "b" or "4" side?

$$4 \times 15 = 60$$

The side that measures 60" is our "4" side. The only side left in our 3-4-5 triangle is the "5" side. 5 x 15 = 75, so the diagonal of the rectangle above is 75". Is that what you got? Try a few on your own to make sure you fully understand the Pythagorean Theorem. It's kind of a big deal in math.

Name: ______________________________ Date: ______________

CHAPTER 3 REVIEW TEST

1. Find the length of the hypotenuse.

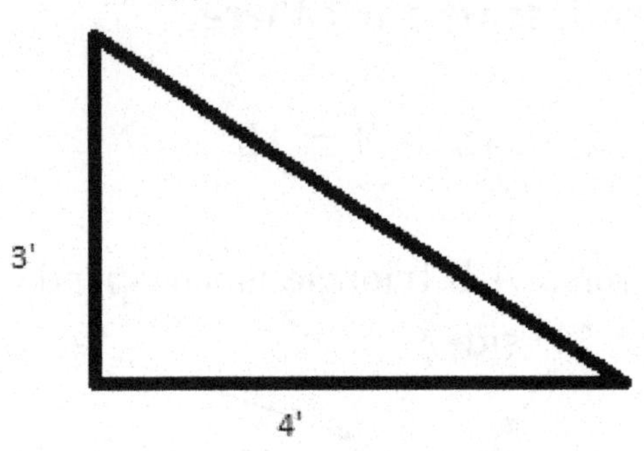

2. Find the length of the hypotenuse.

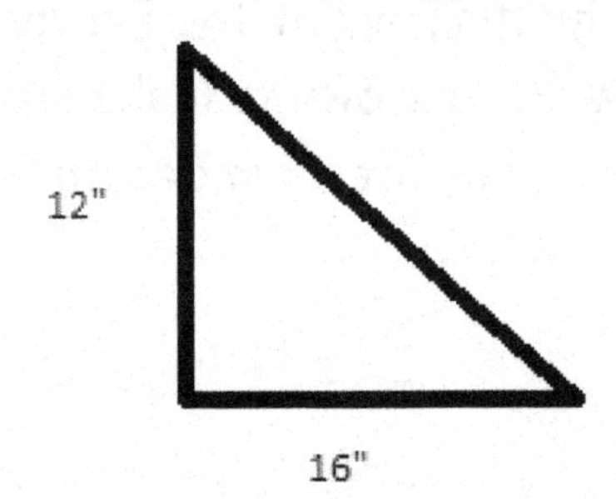

3. Find the diagonal length of this rectangle.

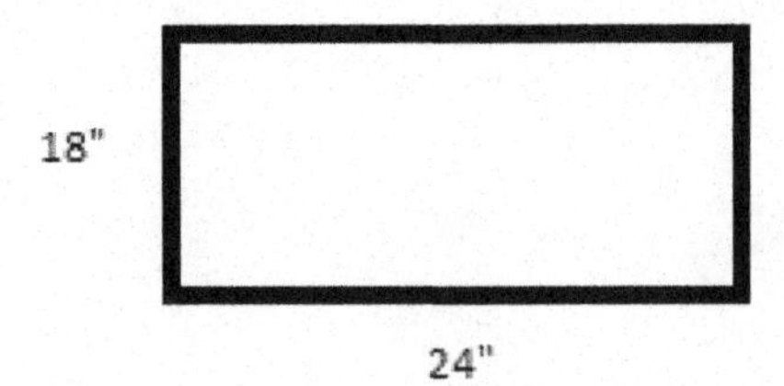

Name: ______________________________ Date: ______________

Chapter 3 Review Test page 2

4. You are told to draw the chalk lines for a football game. The rectangle for the football game is 100 yards long by 25 yards wide. Find the diagonal of this rectangle in order to have a perfectly "square" rectangle. What is the diagonal measurement of the football field?

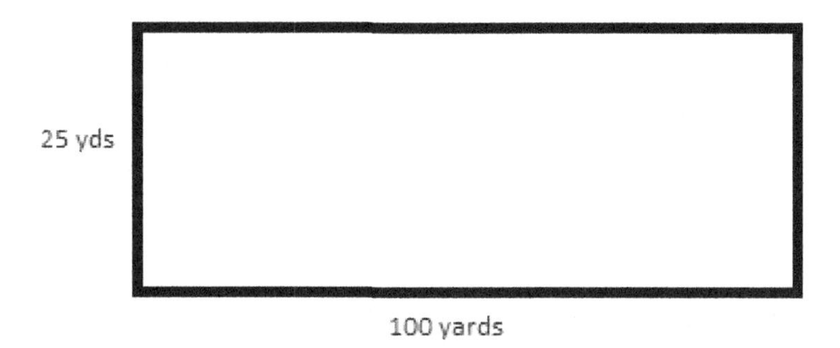

5. Write the Pythagorean Theorem.

6. What is the missing measurement?

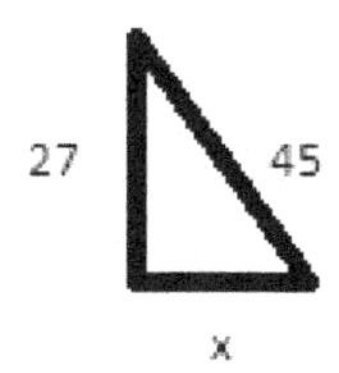

If you understand everything up to this point, you are ready to learn about circles. If you are confused by anything we've learned so far, please go back and read this section again until it is simple.

CHAPTER 4

CIRCLES

LESSON 11: DIAMETER AND CIRCUMFERENCE

Now let's talk about circles. A circle isn't a polygon because it doesn't have any sides. A circle will always measure 360°. It doesn't matter if the circle is as small as a dime or as large as a stadium, it will always measure 360°. But, if you were to walk around a dime or a stadium, the distance you walked would certainly be different. The distance AROUND a circle is called the *circumference.* Of course, not every circle has the same distance around. The bigger the circle the larger the circumference.

Look at the circle below. It measures 1 inch from one side to the other (not exactly, but it's close). That distance is called the *diameter* of the circle. The diameter of the circle below is 1 inch.

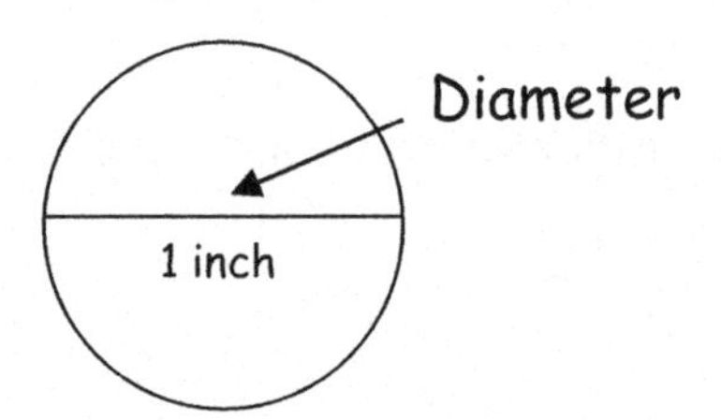

To measure *around* a circle, is to measure the *circumference* of the circle. Look in your Geometry Kit for the two cards that have a wooden circle and a string attached. Start with the smaller circle. It has a diameter of 1". Wrap the string around the circle and then use the ruler to measure the circumference of the 1" circle.

The string should measure somewhere between $3\ \frac{1}{8}$" and $3\ \frac{1}{4}$" because for every one-inch in diameter, a circle will always be 3.14" around.

If the circle were 1 mile in diameter, the circumference would 3.14 miles. Can you guess what the circumference of a circle would be, if the diameter were 1 foot? Of course, it would be 3.14 feet.

The number 3.14 is used so often in geometry, that it has a special name and symbol. The symbol we use to represent 3.14 is π. It is actually a Greek letter called pi (pie), so that's what we call it - pi.

Look at the 2" circle with a string from your Geometry Kit. It is 2" in diameter. Wrap the string around this circle. Since this circle has a diameter of 2 inches, the circumference will be 3.14 x 2. You could also say the diameter is pi times 2 or 2π. Measure the string. It should be just barely over $6\frac{1}{4}$" long; 6.28" to be exact.

That brings us to the next formula on your formula card.

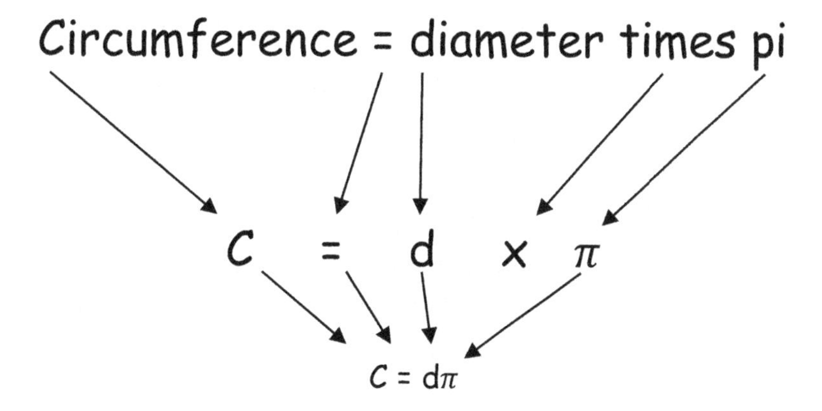

Here is a trick to help you remember how to get the "C"ircumference of a circle...

$C = d\pi$

Seedy pie

You aren't expected to memorize all the formulas, but circumference and area of a circle are used so often, it is helpful to memorize those two.

Try a few on your own. Find the circumference of the following circles.

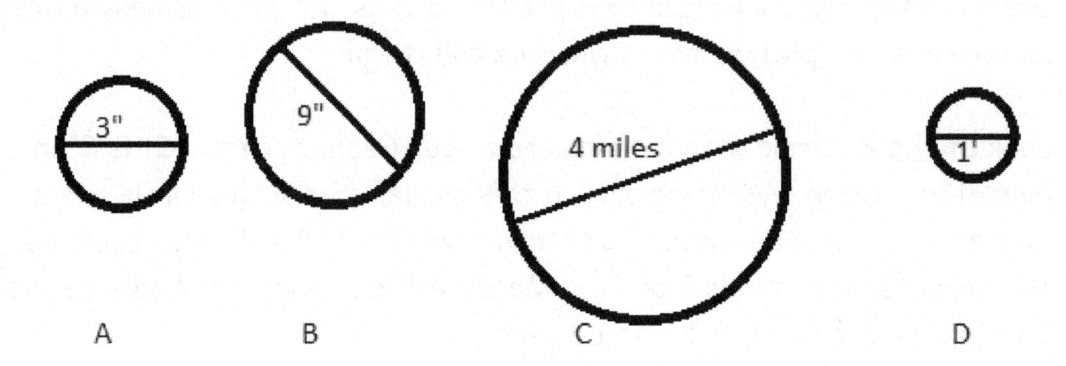

Solutions:
Circle A has a diameter of 3". Using Seedy Pie, we can calculate the circumference by multiplying 3" x 3.14 = 9.42 inches. NOT square inches because we didn't multiply inches x inches, we multiplied inches times pi. That gives us the length around the circle, not the area of the circle.

Circle B has a diameter of 9". Using $C = d\pi$, we can multiply 9" x 3.14. The circumference of circle B is 28.26 inches.

Circle C has a diameter of 4 miles. To find the circumference, we use "Seedy Pie" or $C = d\pi$. C = 4 miles x 3.14 = 12.56 miles around.

Circle D is easy. We don't need to do any math because the diameter is 1'. We already know that for every 1 foot there are 3.14 feet around the circle, so the circumference of circle D is 3.14 feet.

Does that make sense to you? Great! Complete the next worksheet.

Name: ______________________________ Date: ______________

WORKSHEET 4-11

Use $C = d\pi$ (seedy pie) to find the circumference of each circle.

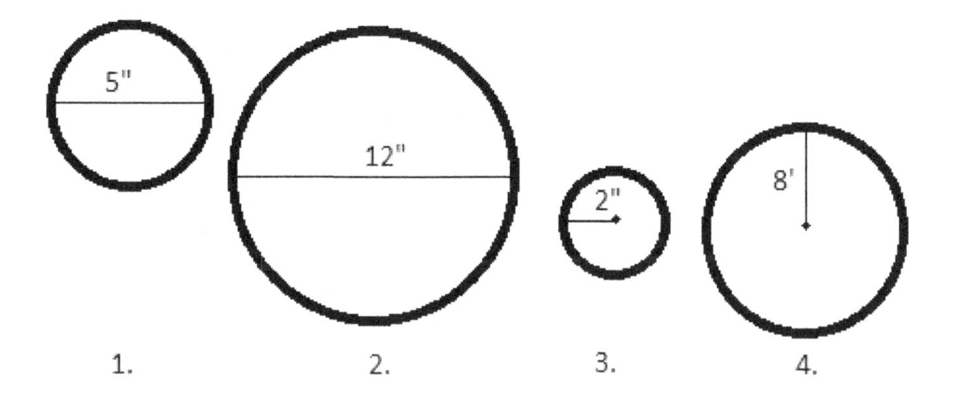

5. Which Greek symbol do we use to represent 3.14 inches?

6. Mick cut down a tree with a chainsaw. The stump measured 6.28 feet around. What is the diameter of the tree he cut down?

7. Josh bought a round swimming pool. It measures 15' from one side to the other side. Josh wants to put a string of Christmas lights around the pool. How long should the string of lights be?

8. Amanda drove over the bridge from point A to point B. It was a 1 mile drive across the bridge. The circle below represents a lake. How long would she have driven, if she had gone around the lake from point A to point B?

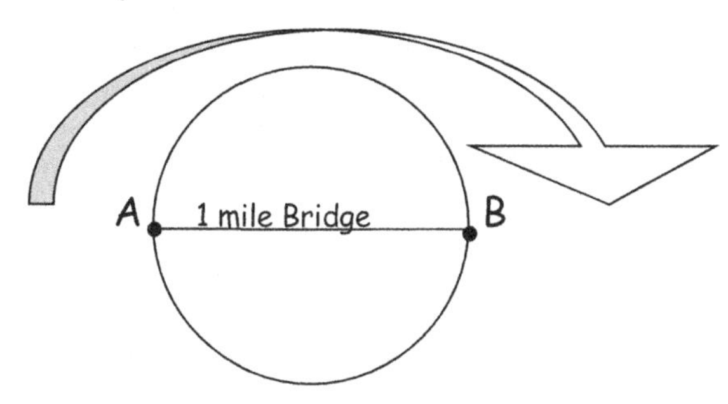

LESSON 12: RADIUS AND AREA

You learned what the diameter of a circle is; the distance from one side of the circle to the other side. Now you will learn about the distance from the center of the circle to the edge.

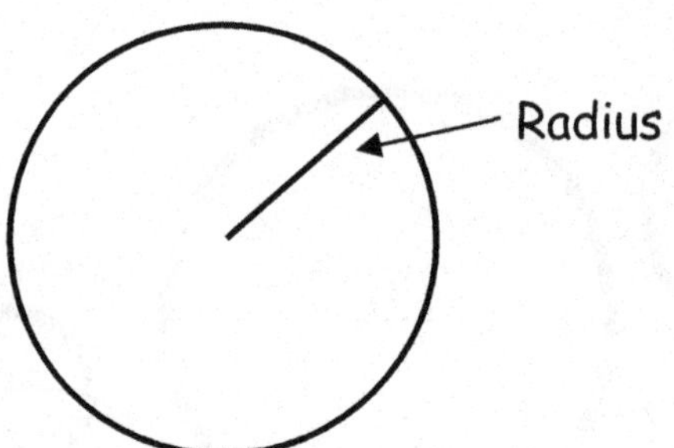

This distance is called the *radius* (ray-dee-us). It's just like a ray whose endpoint is the center of a circle.

Next we will learn how to find the *area* of a circle. Remember, area is like an area rug. We want to find out how big a rug we need to cover the area of the circle. To find area, we will have to use pi again. Here is the next formula on your card.

$$A = \pi r^2$$

Area = pi x radius squared

Let's find the area of this circle using the formula $A = \pi r^2$.

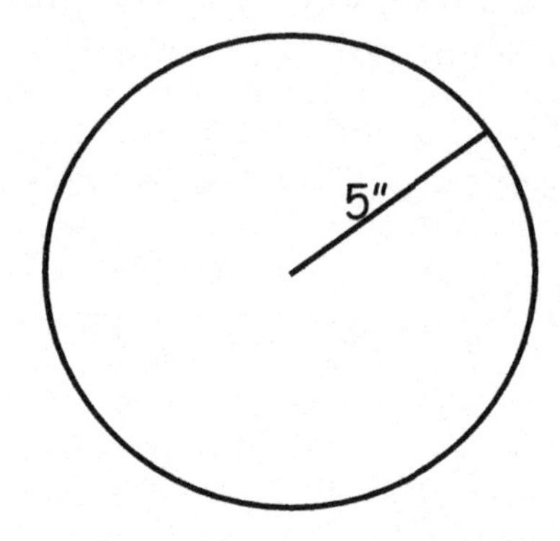

We know the formula, so all we need to do is fill in the blanks.

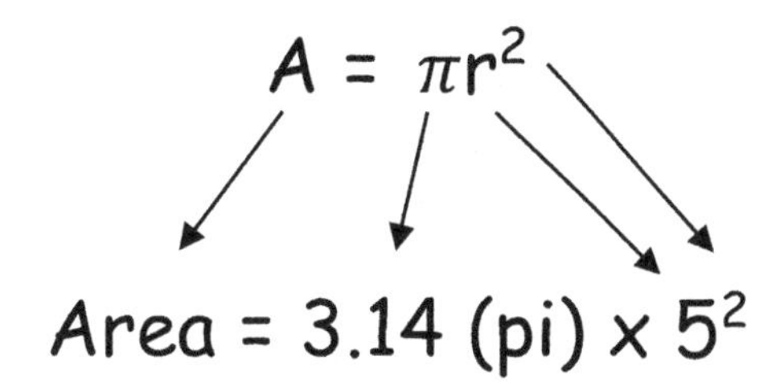

The radius is 5.
5^2 means 5 x 5 = 25.
Pi = 3.14.
Area = 3.14 x 25
Area = 78.5 square inches.

This time the answer *is* in "square inches" because we squared the radius. The formula has a square^2 in it, the radius is squared, and the answer is in square inches. These are all clues that we are finding area not a length. Trick question: The next picture is a circle with a diameter of 10 feet. Can you figure out the length of the radius of this circle?

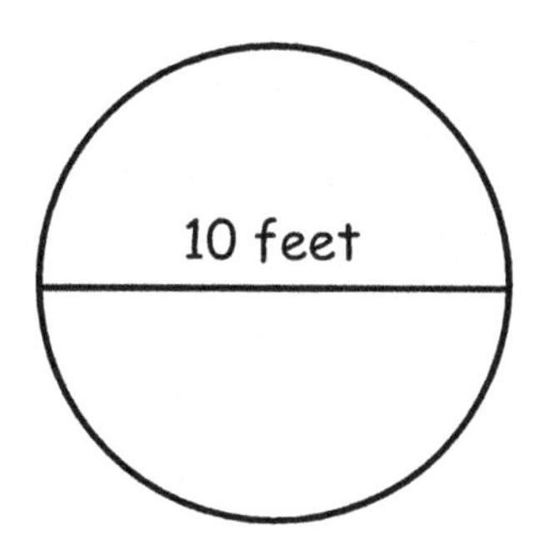

The radius is half of the diameter, so the radius for this circle is 5 feet. With this information, we can figure out the circumference and the area of this circle. Let's figure out the circumference first. Circumference starts with the letter "C," so automatically I know "Seedy Pie" is the formula.

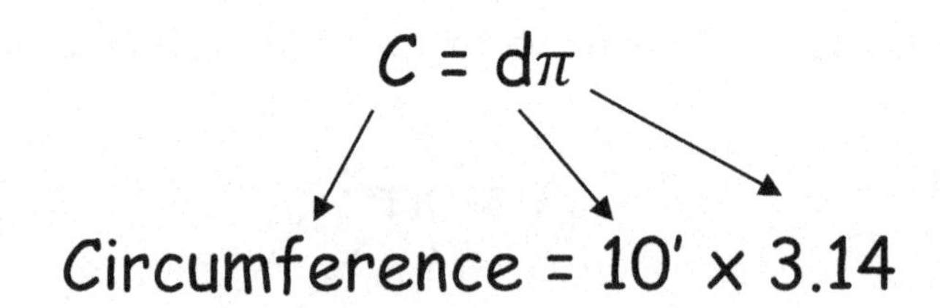

$$C = 31.40 \text{ ft.}$$

By multiplying the diameter by 3.14, I can figure out the circumference. Next, we will calculate the area of the circle. The formula is $A = \pi r^2$, which brings me to a question. Is an apple pie a circle? No silly, a pie are squared. Get it? A pie are squared…A πr^2. That's my little trick for remembering the formula for area of a circle.

Again, you don't have to memorize all the formulas, but a real mathematician can always rattle off the formulas for area and circumference of a circle. And with just two silly phrases, you can too. So remember: "Seedy pie" and "A pie are square" with "C" and "A" being the clues, to which formula is which.

But let's get back to our circle with a diameter of 10′. Let's figure out the area. The formula is "A pie are squared" $A = \pi r^2$. If the diameter is 10′, then the radius is 5′. Let's fill in the formula.

$$A = \pi r^2$$

$$Area = 3.14 \times 5^2$$

$$A = 3.14 \times 25$$

$$A = 78.5\ sq.ft.$$

Name: ______________________________ Date: ______________

WORKSHEET 4-12

1. Find the circumference of a circle with a diameter of 3". Remember to use "seedy pie".

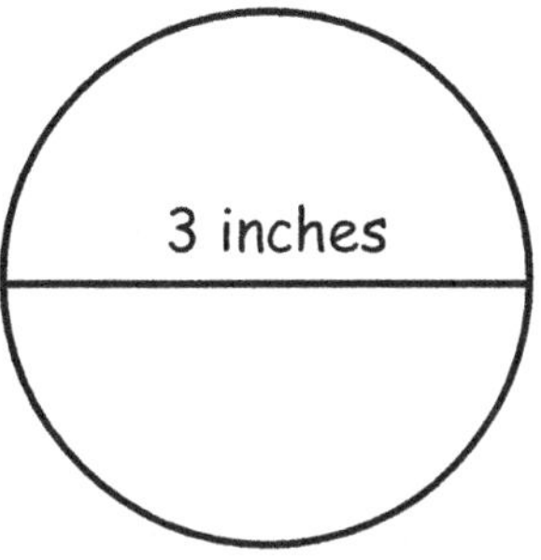

2. Find the area of a circle with a *diameter* of 8". Use "A pie are squared" to figure out the answer and be sure your answer has "inches squared" at the end.

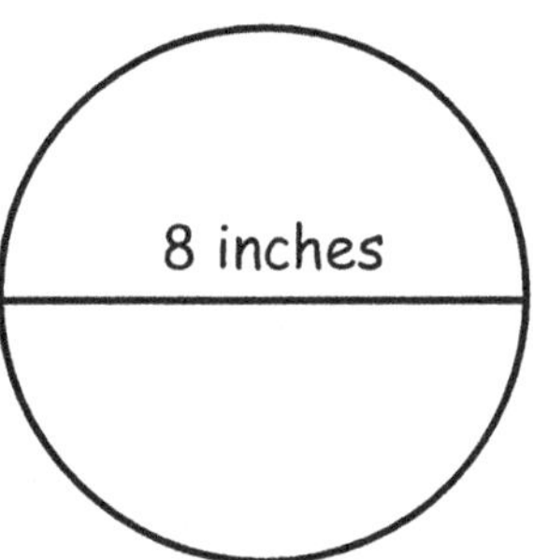

3. Answer this question immediately, don't do any math, just figure it out logically in your head. What is the circumference of a circle whose diameter is 1 inch?

4. What is the radius of a circle that has a circumference of 3.14 inches?

LESSON 13: AREA OF ODD SHAPES

If you understand everything in this book so far, keep reading. If you are even a little bit confused, please go back to where you weren't confused.

If you are able to find the area of a circle, a square, and a triangle, then you are able to find the area of shapes like this next one.

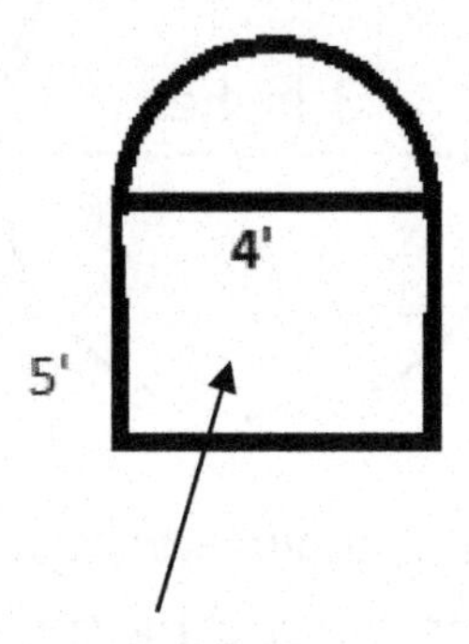

First, find the area of the rectangle, 5' x 4' = 20 square feet.
Next, find the area of a circle with a diameter of 4 and then cut that area in half, since it is only a half circle.

Using "A pie are square" ($A = \pi r^2$) to find the area of a circle, we multiply the radius squared by pi. $2^2 \times 3.14 = 12.56$ square feet for the whole circle. Half of that circle is 6.28 square feet. Add those two answers together and you have the area of this shape. 20 sq. ft + 6.28 sq. ft = 26.28 square feet.

Here is another oddball shape whose area you can find easily.

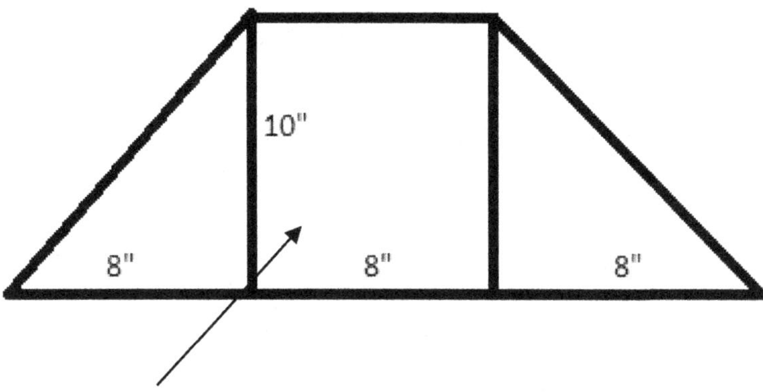

Start with the rectangular shape. 8" x 10" = 80 sq. in.
Next, find the area of the left triangle. 8" x 10" = 80 sq. inches. Cut that number in half and the area of the triangle on the left is 40 square inches. The triangle on the right is the same size, so it too will have an area of 40 square inches. Add these 3 areas together, 80 + 40 + 40 = 160 sq. in.

That sounds simple enough. If you agree, complete the next worksheet.

Name: ______________________________ Date: ______________

WORKSHEET 4-13

1. What is the area of this shape?

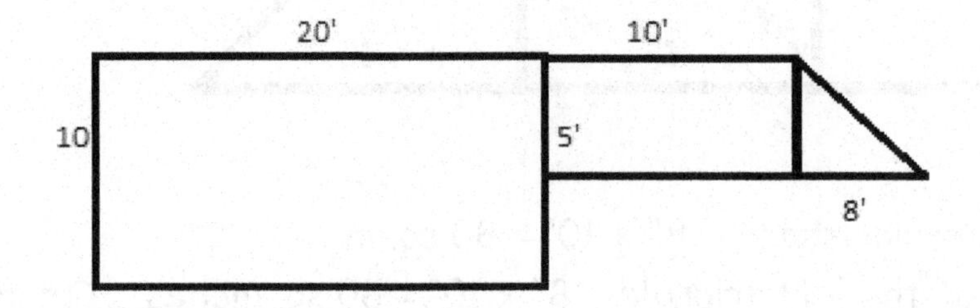

2. What is the area of this shape?

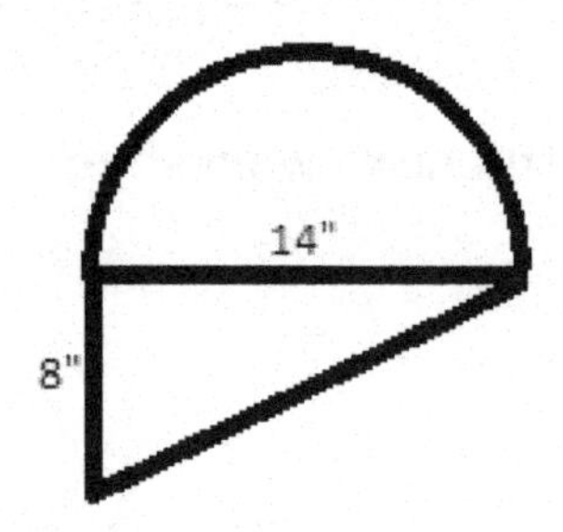

3. Find the area of the gray and white shape below.

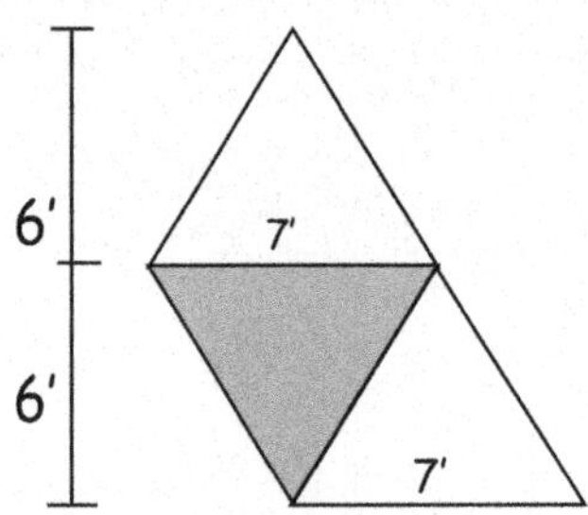

You learned that the *area* of a shape is like an area rug; it covers a flat shape. You learned that *circumference* is the measurement around a flat circle. Now you will learn what *perimeter* means. First, let me tell you a story.

Have you heard about the dog named *Peri*? *Peri* was a dog who loved to chase cats. *Peri's* owners didn't like him chasing cats, so they put up several *meters* of fencing in the shape of a square. If you measure the fence, you would be measuring the *perimeter* of a shape. How many *meters* long is *Peri's* fence? Find the *perimeter* of *Peri's* fence. (Get it? Peri-meter).

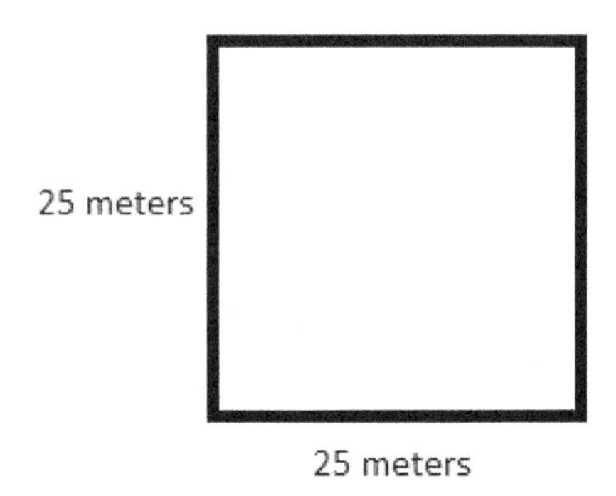

To find the perimeter of the shape above, add up all four sides. Each side is 25 meters, so 25 + 25 + 25 +25 = 100 meters. (Not square meters, that's for area). The perimeter of Peri's fence is 100 meters. Now find the perimeter of the equilateral triangle in this next picture.

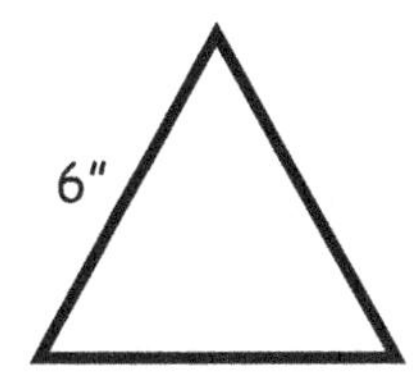

What do we know about this triangle? We know that one side is 6 inches, and we know that it is an equilateral triangle. Equilateral means all sides are equal, so all sides must be 6 inches. Add the three sides together and you get 6 + 6 + 6 = 18. The perimeter of the triangle above is 18 inches. Complete the next worksheet to make sure you understand the meaning of perimeter and circumference.

Name: ______________________ Date: ____________

WORKSHEET 4-14

1. Find the **perimeter** of the following three shapes: a square, a right triangle and a rectangle.

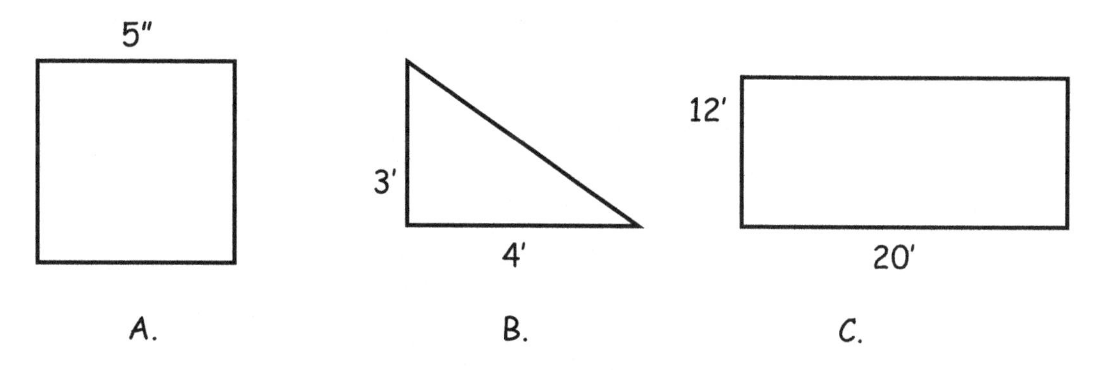

2. Find the circumference of this circle.

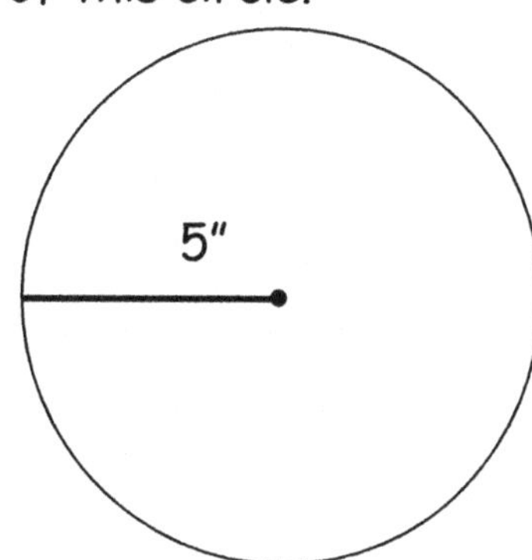

3. Find the perimeter of a square with 3/8" sides.

If you fully understand the meaning of area, perimeter, and circumference, you are ready to learn about volume.

A cube is different from a square because a square is a flat plane figure. A cube is not a plane figure. It has 3 dimensions, so it is called a *space figure.* It has *space* inside that can hold water, sand, or air. It is also called a 3-D shape. You can print a 3-D cube cut-out from our website, LearnMathFastBooks.com.

We can measure how much space there is inside of a space figure using formulas. Measuring the space inside is called *finding the volume.*

A "cube" has all equal sides; just like a square. If the sides are different lengths, it is not really a cube; it is called a rectangular prism, instead.

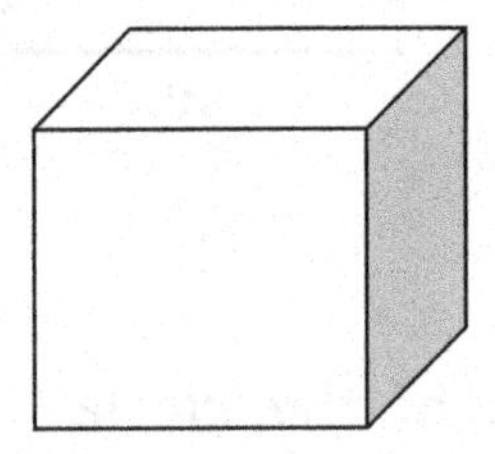

A big jug can hold a gallon of milk, but our little cube cannot hold that much milk. It doesn't have that much *volume.*

In order to find the volume of a cube, we need to multiply **3** numbers. The base, times the height, times the width. Look at the rectangular prism below.

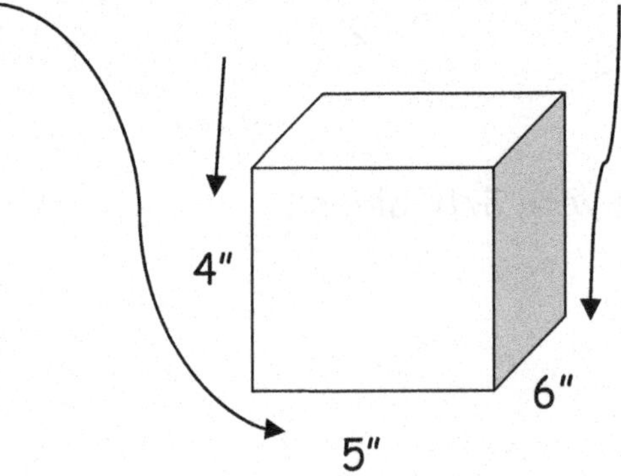

To find the volume, we use the formula V = bhw.

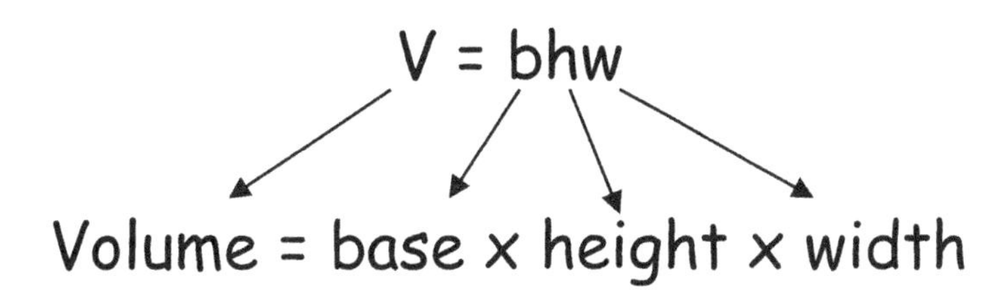

Here is the math for this shape, 5″ x 4″ x 6″ = 120 cubic inches (also written as $120in^3$ or 120 cu. in.). Can you guess why the answer is in cubic inches? These aren't regular inches, they are *cubic inches*. Each inch is measured in a cube because we are measuring space. When you multiply 3 x 3 x 3, the answer can be written as 3^3 or said, "three cubed." Now do you understand why the answer is in cubic inches?

When you figure out the *area* of a square, you multiply 2 numbers and the answer will be in square inches. 3 x 3 can be written as 3^2 or three squared. Do you see how the answer to a volume question is always cubed and the answer to an area question is always squared?

Find the volume of this shape.

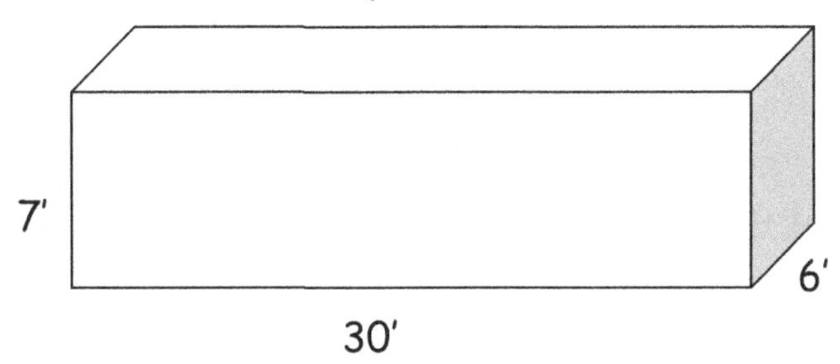

Use the formula V = bhw. Multiply the three numbers in any order you choose. 7′ x 6′ x 30′ = 1260 $feet^3$. Notice, the answer is cubed, since we multiplied feet x feet x feet. You must fully understand this point. When we multiply all **3** dimensions of our **cube,** the answer is in **cubic** feet. It can also be written as ft^3 (feet **cubed**).

Try a few on your own on the next worksheet.

Name: ______________________________ Date: ______________

WORKSHEET 4-15

Find the volume of these shapes.

1.

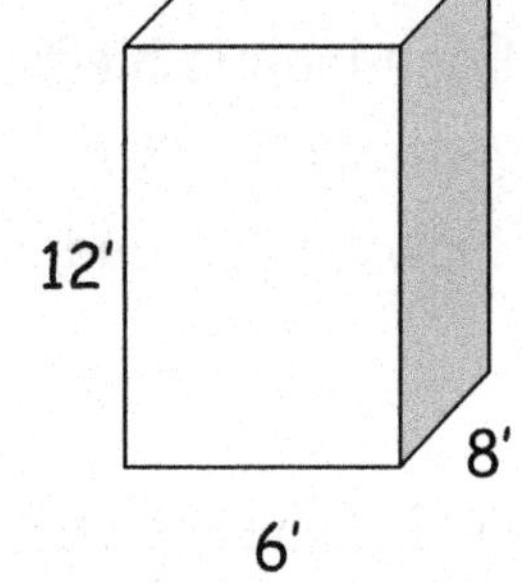

2.

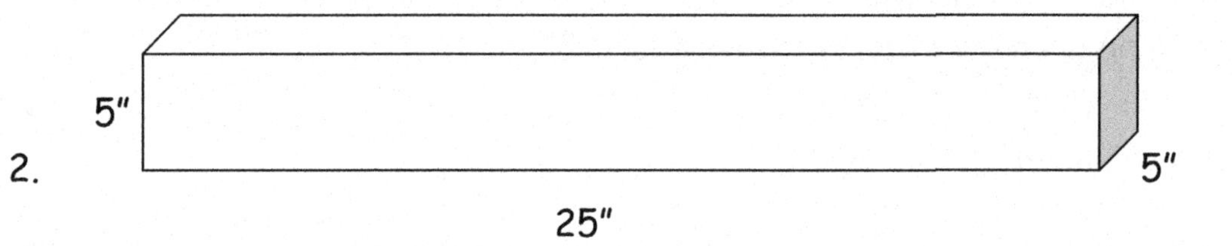

3. How much water can this fish tank hold? Your answer should be in cubic feet.

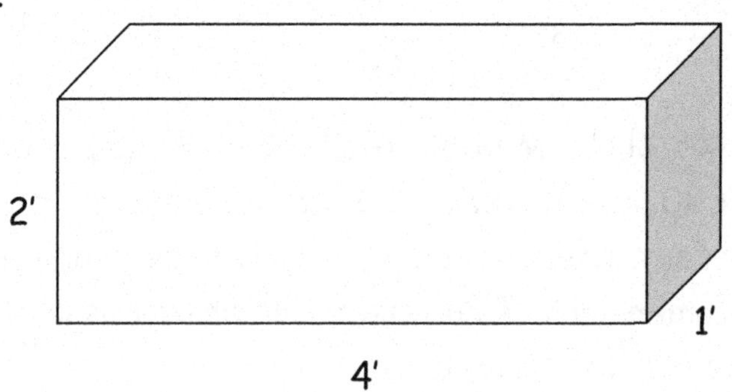

LESSON 16: VOLUME OF A SPHERE

There are many formulas used in geometry to find the volume of different shapes. Some are very long, but they all work the same; just fill in the letters with the numbers you have. You aren't expected to memorize all the different formulas, but you are expected to know how to use a formula.

Let's go over one of the more difficult formulas together. Look in your Geometry Kit for the bouncy ball. If you don't have the kit, find a ball around the house. A ball is not a flat circle. It is called a *sphere*. A sphere is a circular 3-D figure that has space inside of it. To find out how much space is inside of a ball or globe, we use the Volume of a Sphere formula.

$$\text{Volume of a Sphere} = \frac{4}{3}\pi r^3$$

This one looks complicated, but don't worry it will be easy. You know that the radius of a circle is the distance from the center to the edge of the circle. The radius of a sphere is the same thing. It is the distance from the center of the sphere to the outside edge. That is the only number we need to plug into our formula. We will find the volume of a sphere with a radius of 5". Let's fill in the formula.

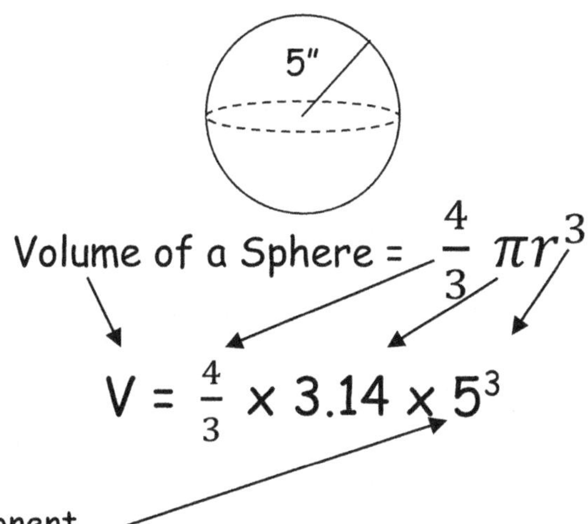

First solve the exponent.

$$V = \frac{4}{3} \times 3.14 \times 125in^3$$

To multiply a fraction by a whole number, put the whole number over 1. The number for pi is not a whole number. Turn it into a mixed number and then into an improper fraction, so we can multiply.

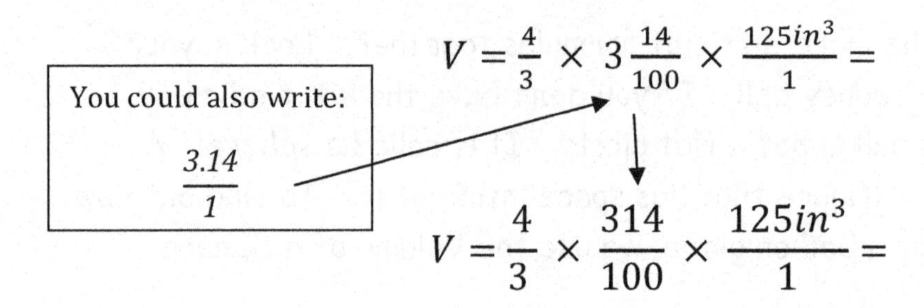

First, multiply all the numerators. $4 \times 314 \times 125in^3 = 157{,}000in^3$
Next multiply all the denominators. $3 \times 100 \times 1 = 300$
Those two answers are your new numerator and denominator.

$$V = \frac{157000}{300} in^3$$

Cross off two zeros from the numerator and denominator.

$$V = \frac{1570\cancel{00}}{3\cancel{00}} in^3$$

$$V = \frac{1570}{3} in^3$$

That fraction means $1570in^3 \div 3$. Do the math.

The Volume of the Sphere is $523.\overline{33}$ in^3.

Try to find the volume of a sphere with a radius of 6". Use $V = \frac{4}{3}\pi r^3$.

Here are the steps you take to find the volume of a sphere with a radius of 6".

1. Write out the formula, but replace the "r" with a "6".
2. Replace the pi symbol with 3.14.
3. Solve for 6^3.
4. Rewrite the formula with numbers.
5. Put any whole numbers over 1, so you can multiply with fractions.
6. Multiply all 3 numerators and put that over the product of all 3 denominators.
7. Do the division. Put in^3 at the end of your answer, since you multiplied 6" 3 times.
8. My answer is 904.32 in^3. Is that what you got?

Practice finding the volume of a sphere, on the next worksheet.

Name: ________________________ Date: ____________

WORKSHEET 4-16

Find the volume of the following spheres.

1. A basketball with a radius of 7".

2. A globe with a radius of 10".

3. A bouncy ball with a radius of $\frac{1}{2}$ inch.

Find the volume of the following rectangular prisms.

4. A box that has a base of 4", stands 5" tall, and has a width of 7".

5. A treasure chest that is 3' wide, 2' tall, and 2' deep.

6. A cube of butter that measures 1" tall, 2.25" long, and 1-1/8" wide.

7. What is a space figure?

8. What is the difference between a cube and a square?

9. What is the difference between a circle and a sphere?

10. What is the difference between a cube and a rectangular prism?

Name: ______________________________ Date: ______________

CHAPTER 4 REVIEW TEST

Find the **C**ircumference of the following circles.

1. A circle with a diameter of 10 yards.

2. A circle with a radius of 7 miles.

Find the **A**rea of the following circles.

3. A circle with a radius of 3 feet.

4. A circle with a diameter of 12 inches.

Find the area of the following shapes.

5.

15 feet

9 feet

12 feet

6. Find the perimeter of the shape above.

Find the volume of the following space figures. $Vol. of\ sphere = \frac{4}{3}\pi r^3$

7.

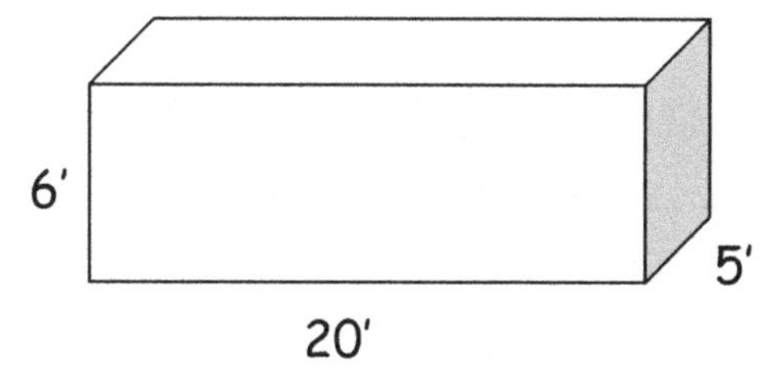

8.

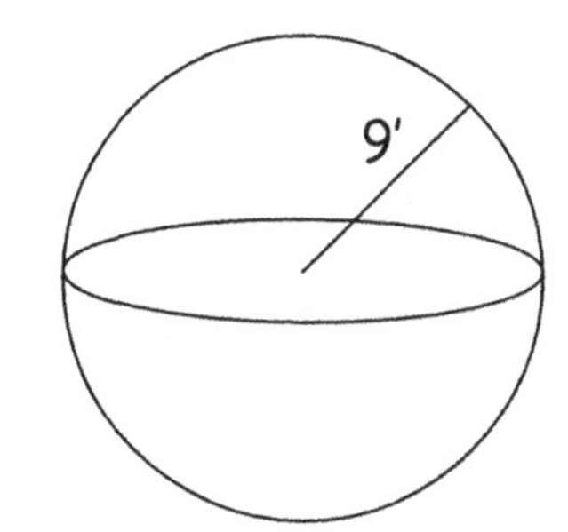

CHAPTER 5

BASIC METRIC SYSTEM

LESSON 17: BASE UNITS

How tall are you? Your answer should be in feet and inches. For example, 5 feet 2 inches tall. That is how we measure length here in The United States; we use feet and inches. When we weigh ourselves, we step on a scale to find out how many pounds we weigh. And when we buy gas, we buy it by the gallon.

Once you leave The United States, you will notice that the other countries don't use inches, pounds, or gallons to measure and weigh things. They use the Metric System instead.

It is completely different from our way. Different words and different sizes, but fortunately it is much easier to learn than our US Standard way.

What makes the metric system so easy to learn is that everything is in increments of 10, just like our money. You know how 10 pennies equal 1 dime and 10 dimes equal 1 dollar? Those are increments of 10, also called Base 10.

With the US Standard way of measuring, 12 inches equal 1 foot, 16 ounces equal 1 pound, and 4 quarts equal 1 gallon. It's not as consistent as the metric system. If you are talking about feet the base is 12, for pounds the base is 16 and for gallons - it's base 4. Luckily, the Metric System is much easier.

The Metric System doesn't measure length in feet; it measures length in *meters*. The Metric System doesn't measure weight in pounds; it measures weight in *grams*. (Actually they call it mass not weight, but you'll learn about that in science class). And instead of measuring *volume* in gallons, the

Metric System uses *liters* (also spelt litres). *Volume* is a measurement of liquid or space.

The base units in the US are: feet, gallons, and pounds. The Metric System uses meters, liters, and grams as base units. When you tell someone your age, you use years to measure time; not days or hours. That's what a base unit is; a set amount that we count as 1, just like 1 year. A year equals 365 days, but we count in years not days - it's the base unit.

In the metric system, the base unit of measurement for length is *meters,* instead of feet. The base unit for weight is a *gram,* instead of a pound. And for volume, the base unit of measurement is a *liter,* not a gallon.

Those are the first 3 words you must learn: meter, liter, and gram (oh my). Answer the following questions.

In the metric system the base unit of measurement for:

Volume is ________.
Length is ________.
Weight is ________.

That's the hard part, remembering those 3 different base units, so let's talk about them a little more. A **meter** is about 40" or about as long as an adult's leg. A chair is about a meter tall and a door knob is about a meter from the floor.

A **gram** is very small, not even close to a US pound. A paperclip weighs about 1 gram. A piece of typing paper weighs about 5 grams.

Have you ever bought a 2 **liter** bottle of pop? It usually comes in a tall plastic bottle and you can pour around 8 glasses of pop from it. That's what 2 liters looks like. Have you ever bought a bottle of water? The typical size sold is a half-liter of water.

There is one other size bottle of water or pop you can buy. It is smaller than a 2 liter bottle and bigger than your typical bottle of water. It is a 1 liter bottle. Can you picture one of those 1 liter bottles? Now you can picture what 1 liter of liquid looks like.

Name: ________________________________ Date: ______________

WORKSHEET 4-17

Fill in the blanks with the appropriate unit of measurement. Answer with meters, liters, or grams.

1. The golf ball traveled 200 _____________.

2. The letter only weighs 20 __________, so 1 stamp is enough postage.

3. A bottle of water measures 1/2 ________________.

4. How many ______________ does a penny weigh?

5. How many ______________ of gas did you put in your car?

6. It took me almost an hour to walk 3000 _____________.

7. Read the back of the box. How many ______ of sugar are in this snack?

8. A bag of chips holds 340 _________ of chips.

9. The fish tank broke and spilt _______ and ______ of water on the floor.

10. Did you see Mick jump? He must have traveled at least 3 ________.

Once you feel confident that you know the difference between a meter, a liter, and a gram, you are ready to continue.

And now for the beauty of the metric system. In America, we have different names for measurements that are bigger or smaller than our base unit. For example, we use inches, yards, and miles for measurements that are bigger or smaller than a foot. And we use ounces, pints, cups, and teaspoons, for amounts smaller than a gallon.

The metric system is different. Instead, they just add a prefix to meter, liter, or gram, to say how many meters, liters, or grams we are talking about. For example, 1000 meters is called 1 kilometer. 1000 grams is called 1 kilogram and 1000 liters is called 1 kiloliter. They just add the prefix "kilo" to each base unit. Here is a list of some more prefixes:

Kilo	1000 times more
Hecto	100 times more
Deka	10 times more
Deci	1/10
Centi	1/100
Milli	1/1000

All you have to do is slap the right prefix in front of the correct base unit to say how many meters, liters, or grams you are measuring. Let's practice with the prefix Kilo. It means 1000 and it is one of the most common prefixes.

1000 grams = 1 Kilogram.
1000 liters = 1____________________.
1000 meters = 1____________________.

The next most common prefix is "milli" and it means 1/1000. It takes 1000 millimeters to equal 1 meter. It takes 1000 milligrams to equal 1 gram and 1000 milliliters is the same as 1 liter. Complete the next worksheet.

Name:______________________________ Date:______________

WORKSHEET 4-18

1. One paper clip weighs about 1 gram. If I break that paper clip into 1000 equal pieces, each piece will weigh 1 ________________.

2. I bought 1 liter of pop. I gave 1000 people a few drops each. Each person received 1 __________________ of pop.

3. I'm trying to measure how thin my credit card is. It is hard to measure something so small, but I know if I stack up 1000 credit cards, it will stand 1 meter tall. What is the thickness of just 1 credit card? _____________.

4. Sarah weighed herself. She weighed over 200 ___________.

5. Jack filled up the swimming pool. It took over 3 _________ of water to fill it.

6. The pharmacist said these pills have 100 __________ of vitamin C.

7. I stacked up 10 dimes. The whole stack measured 10 ________ tall.

8. The speed limit on the highway is 80 ___________ per hour in Canada.

9. The 10-K race was 10 ______________ long.

10. The recipe calls for 5 _____________ of oil.

11. That bottle has 120 _____________ of perfume in it.

12. 1 gram x 1000 = 1 ______________,

13. 1 gram ÷ 1000 = 1 ______________.

Let's review what you've learned, so far. The 3 base units of measurement in the metric system are liter, meter, and gram. The 2 prefixes you've learned are "kilo" and "milli." Kilo means 1000 times bigger, and milli means 1000 times smaller. That is really all you have learned so far. If you understand this paragraph, you are ready to learn more.

The next most common prefix is "centi." Look at the ruler on your protractor. On the bottom edge is a 6 inch US ruler. Above that is a metric ruler with 10 centimeters. Each centimeter is about the width of a fingernail. The 10 little spaces within each centimeter are millimeters. Each millimeter is about the width of a line drawn with a pencil.

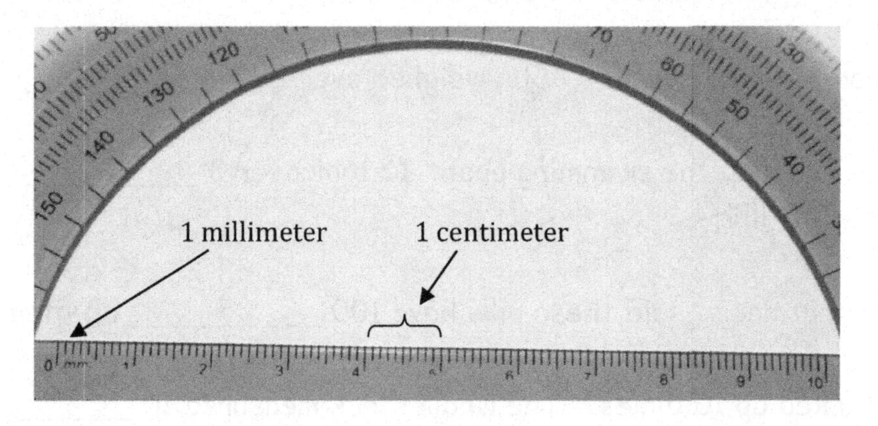

Think about one dollar. How many cents are in a dollar? There are 100 cents in a dollar, right? It's the same with centimeters. 100 **cent**imeters equals 1 meter just like 100 **cents** equals 1 dollar. If I had 100 centiliters, how many liters would I have? If I had 100 centigrams, how many grams would I have?

Can you think of something that might weight about 1 centigram? It would be something very, very lightweight - that's for sure. I would guess that 1 strand of hair might weigh about a centigram.

So far, you have learned 3 of the prefixes for our metric measurements: kilo, milli, and centi. With these 3 prefixes and the 3 base units, you should now know the meaning to the 9 words on the next worksheet. Practice using them, by completing the next worksheet.

Name:______________________________ Date:________________

WORKSHEET 4-19

Use one of the words from the list below to complete these sentences.

Kilogram	= 1000 grams	Centigram	= 1/100 of a gram
Kilometer	= 1000 meters	Centimeter	= 1/100 of a meter
Kiloliter	= 1000 liters	Centiliter	= 1/100 of a liter

Milligram	= 1/1000 of a gram
Millimeter	= 1/1000 of a meter
Milliliter	= 1/1000 of a liter

1. I got a splinter in my hand. It was about 1 __________ wide.

2. I drove to the store. It was about 3 _________________ away.

3. That fish tank is huge. It must have nearly 100 ________________ of water in it.

4. My doctor said this pill has 500 _______________ of aspirin in it.

5. A small bird feather weighs approximately 7 ______________.

6. I went on a diet and I lost over 9 __________________.

7. I put 1 spoonful of water in the dough, that's about 1 _____________.

8. The diameter of a marble is about 1 ________________.

9. A bottle of eye drops holds 30 milliliters, that's the same as 3 ________________.

10. I ran 5 __________ in 1 hour.

Double check your work, to make sure your answers are right. If this was difficult for you, or if you *guessed* some of the answers, read this entire chapter over again.

For more practice, read the labels on items from your kitchen. For example, get a loaf of bread and find where they print the Net Weight of the bread. It will probably be written in US Standard weight and in metric units. One of the two will be inside parentheses.

I grabbed a few items from the kitchen. A loaf of bread is 637 grams. A can of tuna fish has 141 grams of fish in it. A can of chili holds 425 grams of chili. These items are all solid food, so they measure it in grams.

Look for some liquid items in your kitchen. Look for a can of pop or a bottle of juice. These items are liquid, so they are measured in liters. You may see both the US Standard measurement and the Metric measurement. If they are both listed, one of them will probably be in parentheses. The metric units are typically abbreviated.

The bottle I'm looking at writes it like this:

1.75 L (1 QT 1 PT 11 FL OZ)

This means the bottle has 1.75 liters of juice in it. The US Standard measurement is written in the parentheses. That means 1.75 liters = 1 quart, 1 pint, and 11 fluid ounces. Metric units are often abbreviated. The base units are always abbreviated like this:

Meter = m gram = g Liter = L

Pretty obvious, huh? But notice that *liter* is abbreviated with a capital L. Meter and gram are both written in lower case letters.

The prefixes also use their first letter as an abbreviation; all in lower case letters. The only exception is the two prefixes that both start with the letter d: deka and deci. "Deka" is abbreviated as *da*, and *deci* is just the letter *d*.

When you pair up one of the prefix abbreviations with any one of the base unit abbreviations, you can create an abbreviation for any metric unit. For example, to get the abbreviation for centimeters, use the first letter of the prefix *centi* and the first letter of the base unit *meters*.

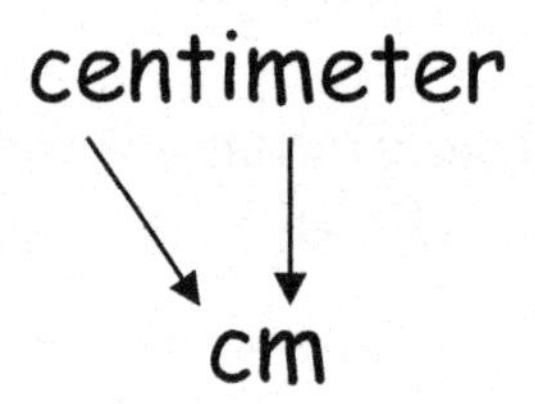

Let's try another one. What is the abbreviation for milliliter? Use the first letter of the prefix milli and the first letter of the base unit. The base unit is liter. Remember that liter is abbreviated with a capital L, so the abbreviation for milliliter is mL. Sometimes it is written with a cursive *ℓ*, so it isn't confused with a capital I. But a capital L is the correct way.

Name: ______________________________ Date: ______________

WORKSHEET 4-20

Write the abbreviation for each of the following metric units.

1. Meter
2. Centimeter
3. Liter
4. Centiliter
5. Kilogram
6. Gram
7. Milligram
8. Kilometer
9. Milliliter
10. Kiloliter

The last prefix we will discuss is *deci*. Deci means 1/10 of the base unit. It takes 10 decimeters, to make 1 meter. The metric ruler on your protractor is 1 decimeter long. If you had 1 liter of pop and you poured a little bit into 10 different glasses, you would have 1 deciliter in each glass. I don't hear people use the word decigram very often, but you should still know that a decigram is 1/10 of a gram.

Think of the prefix deci as a dime. A dime is 1/10 of a dollar, a decimeter is 1/10 of a meter. The prefix "centi" is like a cent. A cent is 1/100 of a dollar, a centimeter is 1/100 of a meter. Get it? Cent...centimeter and dime...decimeter?

There is one more clue to learning these prefixes. Look at this list of prefixes. Each one can make a base unit smaller.

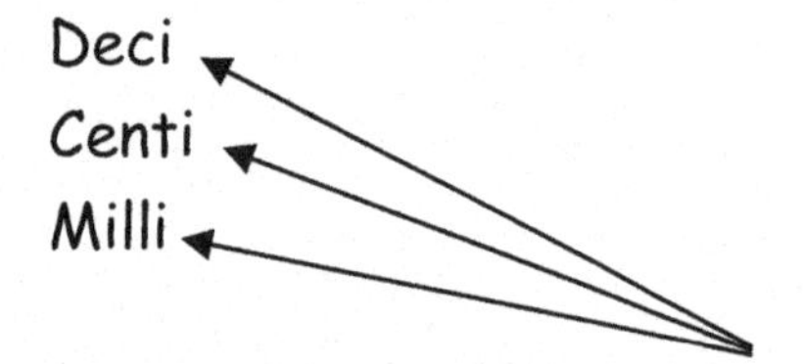

They all end in the letter i. That reminds me of the word "mini," it ends in the letter i too.

Here are the prefixes that make a base unit bigger.

Deka
Hecto
Kilo

They end in the letters o and a, just like the word "mega" or "humungo" (humungo is not a real word, it is slang for humungous). That's how you can quickly recognize if the prefix is making the base unit bigger or smaller. With those clues in mind, let's look at the prefixes deci, centi, and milli one more time.

When you see the word decimeter, you should automatically notice it starts with the letter "d" - just like the word "dime." That is your first clue. It is 10 times bigger or smaller than a meter. Since deci ends in the letter "i," you instantly know it is a "mini" or a tiny amount, so it must be 10 times smaller than a meter.

Now look at the word centigram. The prefix is centi - just like a penny. That is your clue to recognizing that centigram must be 1/100 of a gram. Just like a penny is 1/100 of a dollar.

Look at the metric ruler on your protractor again. The smallest measurement, the space between those little dashes is called a millimeter. You should learn this measurement. A lot of things in America are measured in millimeters. The thickness of a plastic bag is measured in millimeters. Some socket wrenches are sized in millimeters. People use the word millimeter in everyday speech, so you should just learn the word and the size it represents.

You should now be able to distinguish the difference between a millimeter, a centimeter, and a decimeter. If you don't have a clear understanding, read this section again. It won't take long and I'm sure you'll get it the second time.

Below is a metric unit of measurement. Look for clues to help you figure out the meaning of this word. What is the abbreviation?

Kilometer

Here is clue number one. It starts with the prefix "kilo," which ends in the letter "o," so I know it is "humungo" or bigger than a meter. The prefix kilo means 1000, so a kilometer is 1000 meters. The abbreviation is the first letter in kilo and the first letter in the word meter, or km. A kilometer is a little longer than a half mile.

Here is another metric measurement. Look for clues to figure out how big it is and then think of something that would be that size.

Centiliter

The first clue is "centi." That makes me think of 1 cent and since it ends in the letter "i", it must be 100 times smaller than a liter. A liter is about as much as a super large drink at a fast food restaurant. There are 100 centiliters in one liter, so I would guess a centiliter is about a tablespoon or so. The abbreviation, of course, is cL.

If you understand everything we've discussed about the Metric System, then you are ready to take the Chapter Review Test.

Name:________________________________ Date:______________

CHAPTER 5 REVIEW TEST

1. How many centimeters are in a decimeter?

2. How many decimeters are in one meter?

3. How many millimeters are in a centimeter?

4. How many centimeters are in a meter?

5. What measurement is closest to the width of a pencil lead?
   1 mm   1 cm   1 dm   1 m

6. What measurement is closest to the length of a new pencil?
   2 mm   2 cm   2 dm   2 m

7. What measurement is closest to the height of a door?
   2 mm   2 cm   2 dm   2 m

8. What measurement is most likely the size of a chocolate chip cookie?
   7 mm   7 cm   7 dm   7 m

9. What does the prefix Kilo mean?

10. What does the prefix Milli mean?

11. Which is most likely 1 mL?
    A glass of milk   Volume of a fish tank   3 rain drops

12. Which is heavier?
    3 grams   9 centigrams   1 kilogram

Chapter 5 Review Test page 2

13. In the United States, we use feet for our base unit to measure length. What is the base unit used in the Metric System to measure length?

14. Your protractor has a 1 decimeter ruler on it. How many millimeters are in 1 decimeter?

15. Write the abbreviations for the following metric units.

Kilometer
Liter
Hectometer
Dekagram
Meter
Deciliter
Centimeter
Milligram
Gram
Milliliter

If you have completed this book and you feel you are ready to take the final test, begin. If you feel confused, even a little bit, you should go back and read this book again. It won't take long to read it a second time.

FINAL GEOMETRY TEST

1. What is the name of the geometric drawing below?

2. What is the short name for this geometric drawing?

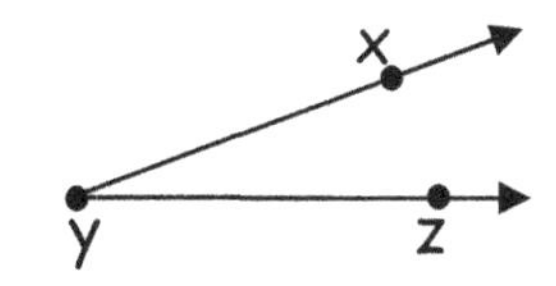

Look at the drawing below and answer the following questions.

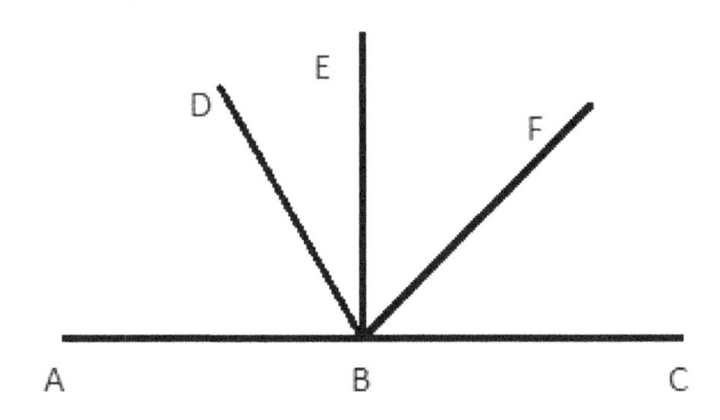

3. Angle EBC is a right angle. What is the measurement of angle EBA?

4. Is angle DBC obtuse or acute?

5. Is angle FBC obtuse or acute?

6. Angle EBC is a right angle. If angle EBD is 25°, then what is the measurement of angle ABD?

7. What angle is adjacent to ABD?

8. If angle FBC is 45° what is the measurement of angle ABF?

9. What angle is adjacent to angle FBC?

10. If angle ABD is 70° what is the measurement of angle DBE?

Final Geometry Test page 2

Look at the picture below and then answer the following questions.

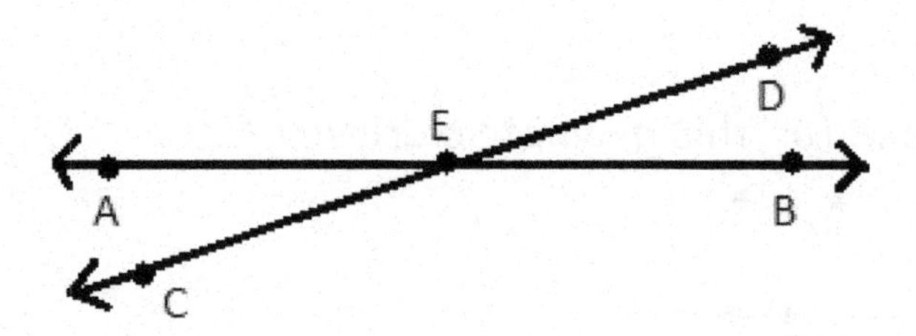

11. What angle is opposite angle AEC?

12. What is the vertex of angle DEB?

13. What angle is opposite angle CEB?

14. Name an angle that is adjacent to angle AED.

15. If $\angle$ AEC = 15°, what is the measurement of $\angle$ DEB?

16. If $\angle$ AEC = 15°, what is the measurement of $\angle$ CEB?

Name each type of special triangle.

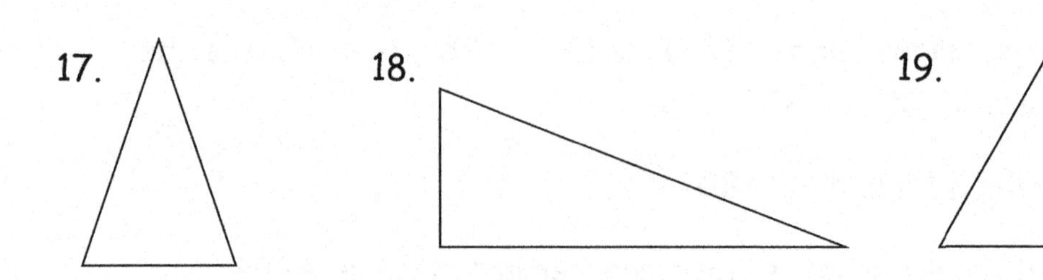

Final Geometry Test page 3

20. Look at triangle ABC below. If $\angle C = 30°$, what is the measurement of angle A?

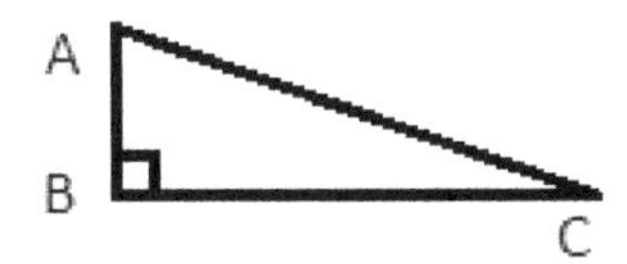

21. Look at triangle ABC below. It is an Isosceles triangle. If angle A = 40°, what are the measurements of angles B and C?

22. Look at triangle ABC below. What is the measurement of angle A?

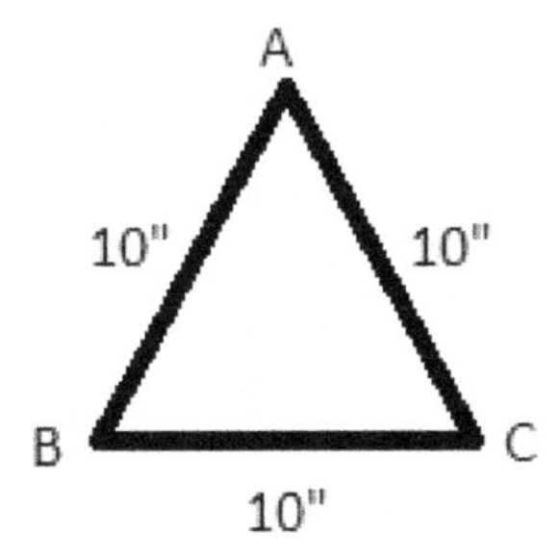

Final Geometry Test page 4

23. Find the area of this triangle.

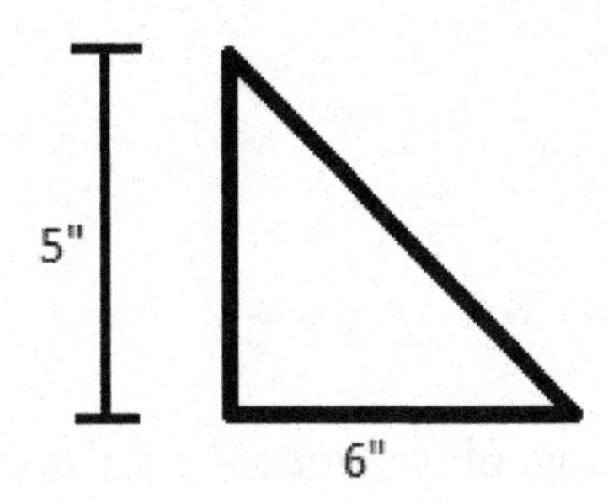

24. Find the area of the rectangle.

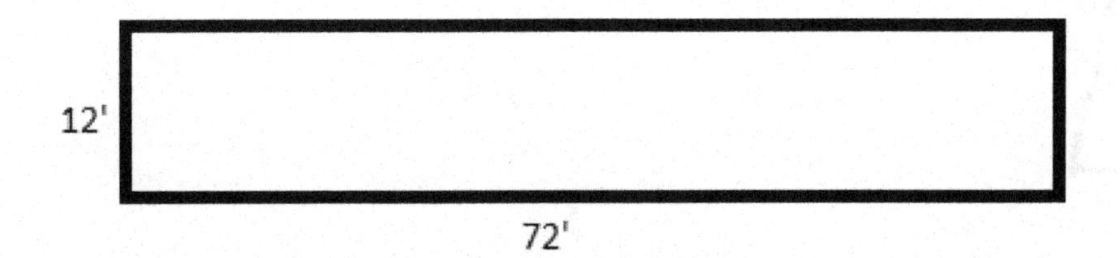

25. Find the area AND the circumference of this circle.

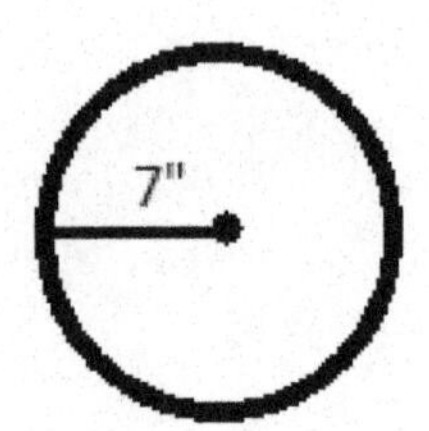

Final Geometry Test page 5

26. Find the area AND the circumference of this circle.

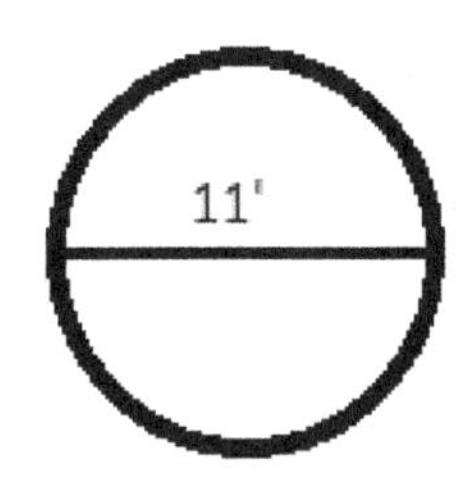

27. Find the perimeter of the shapes below.

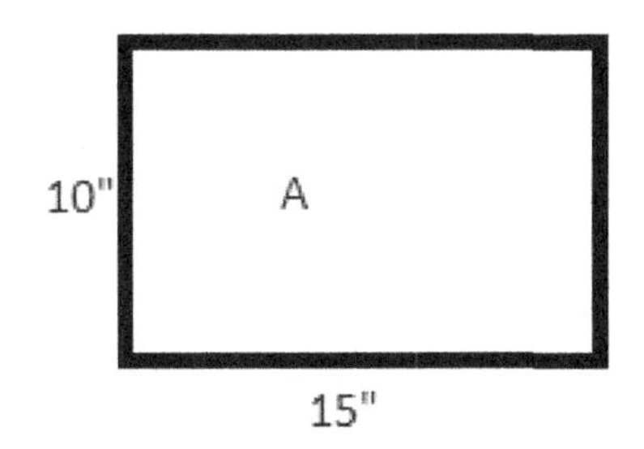

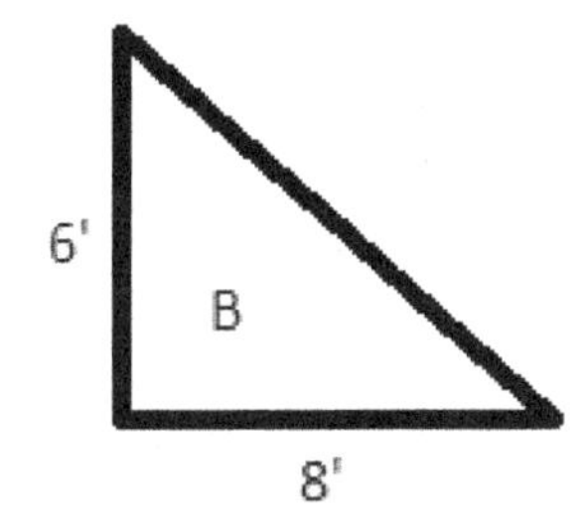

28. Which set of lines are parallel to each other?

A.

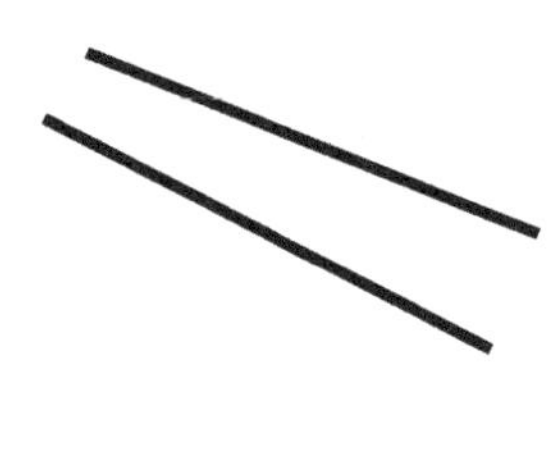

B.

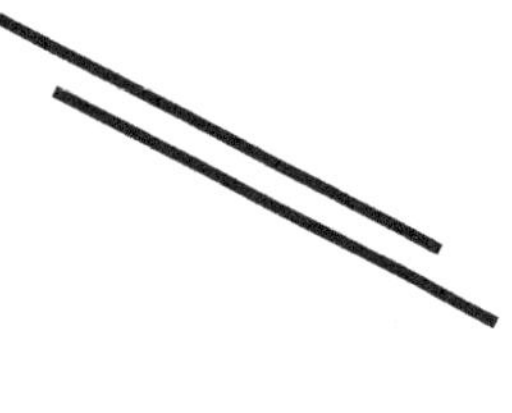

C.

Final Geometry Test page 6

29. Which set of lines are perpendicular to each other?

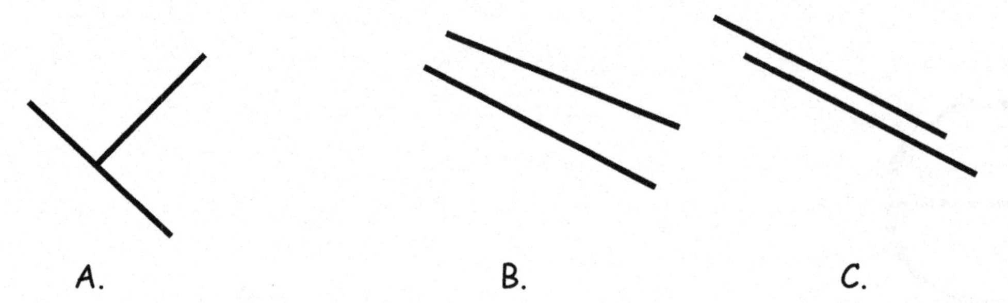

30. Which side of this triangle is the hypotenuse, A, B, or C?

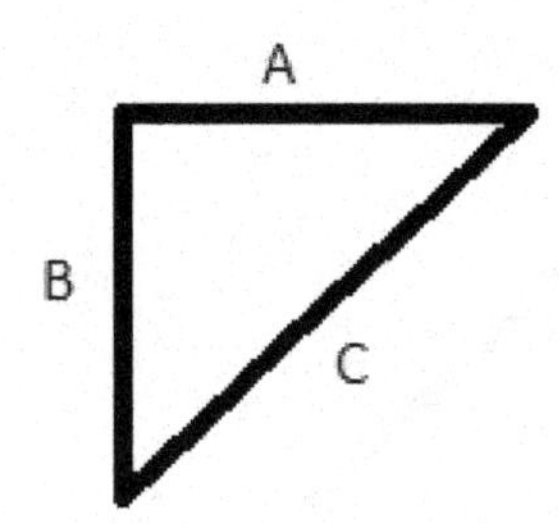

Use the Pythagorean Theorem to find the length of the hypotenuse for the following two triangles.

31.

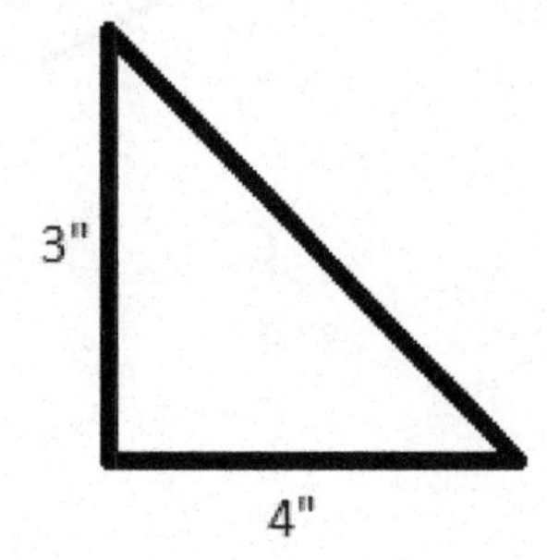

32.

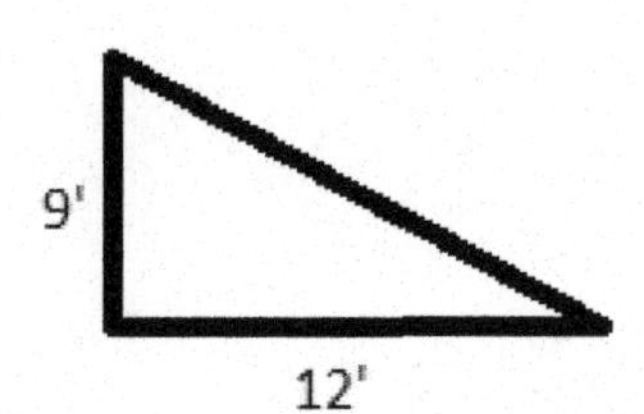

Final Geometry Test page 7

33. What is the diagonal length of this square? Your answer will be the square root of something.

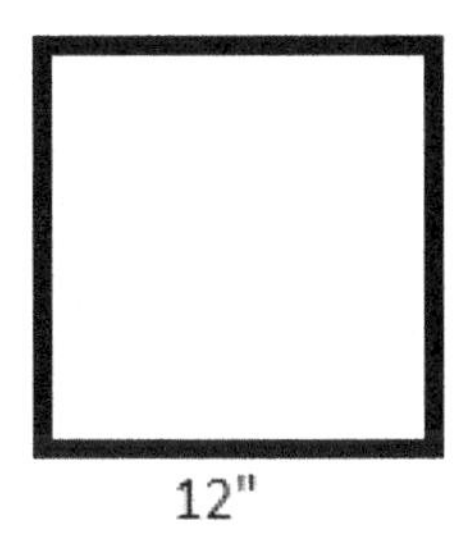

34. Find the area of this oddball shape.

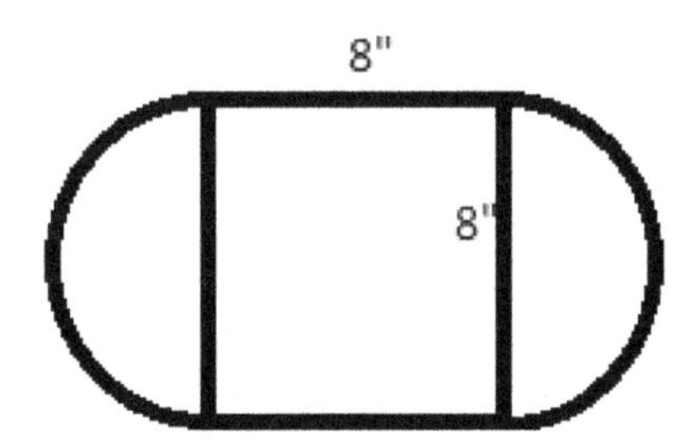

35. Find the volume of the cube below.

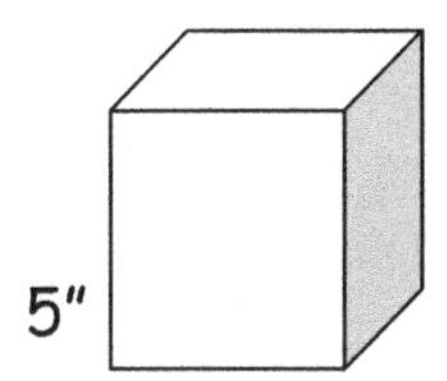

Final Geometry Test page 8

36. Find the volume of the rectangular prism below.

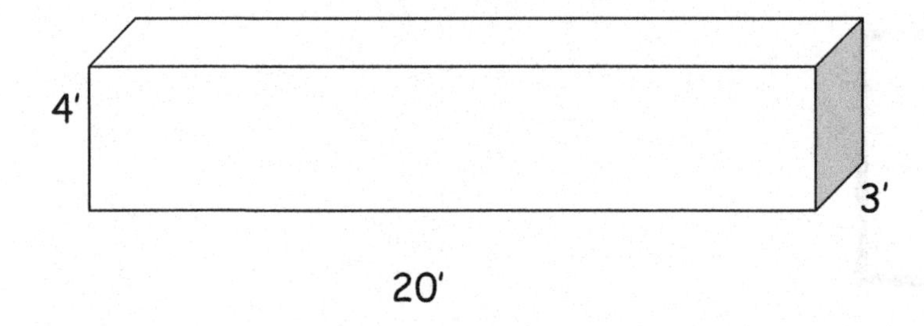

37. Use your protractor to find the measurement of the following angles.

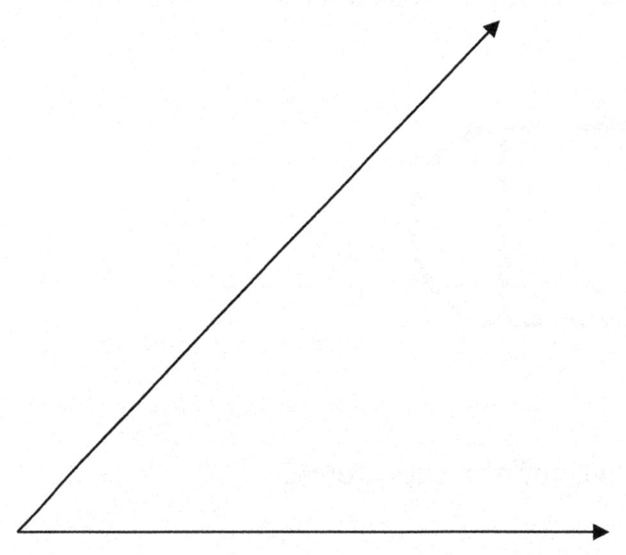

38. Who is on the cover of this book?

39. Find the volume of a sphere with a radius of 7". Use V = $\frac{4}{3}\pi r^3$.
You can round your answer down to the nearest hundredth.

Final Geometry Test page 9

40. What is another name for 1000m in the Metric System?
41. How many milliliters are in 1 liter?
42. The shape below has an area of $54cm^2$. What is the length of base x?

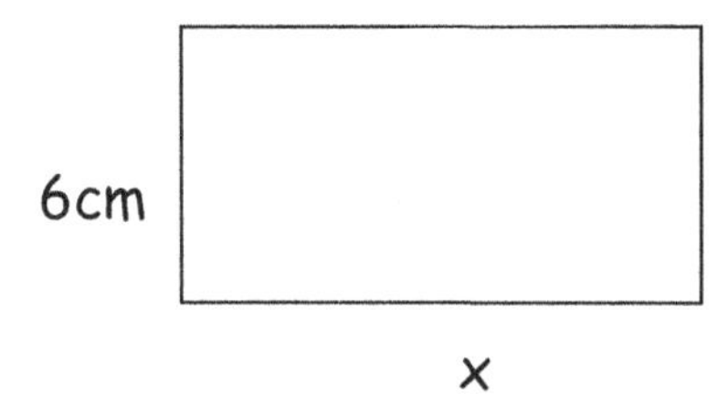

43. How many dimensions are on a plane figure?
44. How many dimensions are on a space figure?
45. Which metric unit of measurement is closest to the US Standard inch?
46. Which metric unit of measurement is closest to the US Standard teaspoon?
47. Which metric unit of measurement is closest to the US Standard yard?
48. Write the Pythagorean Theorem.
49. Which side of the triangle below is the hypotenuse?

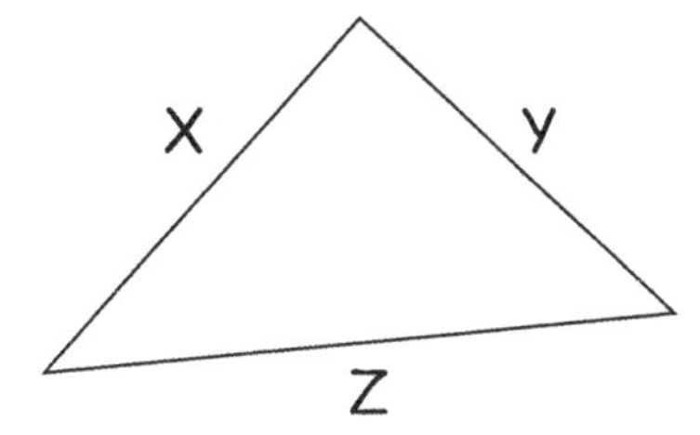

50. What is the name of a 3-sided polygon?
51. What is the name of a 5-sided polygon?
52. Give 3 names for a 4-sided polygon.
53. How many sides does an octagon have?
54. How many sides does a hexagon have?
55. Is a circle a polygon?
56. What formula has the nickname Seedy Pie? What do you use this formula for?

57. What formula has the nickname A Pie Are Squared? What do you use this formula for?

ANSWERS: WORKSHEET 4-1

1. Draw a line with two points creating a line segment AB.

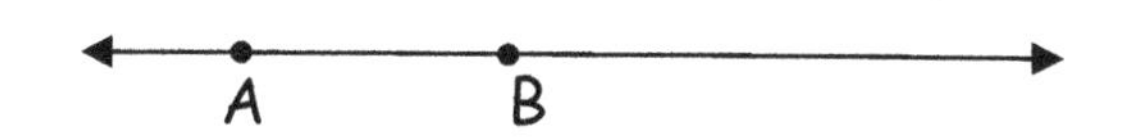

2. Draw an angle and label the points as CDE with point D being the vertex.

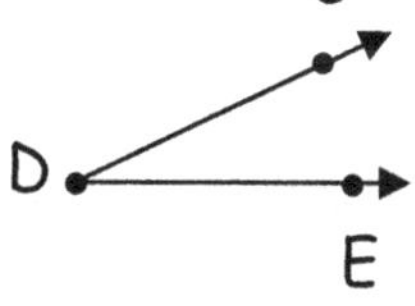

3. Draw ray KL with K as the endpoint.

4. a. Angle
   b. Line Segment
   c. Ray
   d. Line

5. a. Line AB $\overleftrightarrow{AB}$

   b. Line segment CD $\overline{CD}$

   c. Ray BC $\overrightarrow{BC}$

   d. Angle EFG $\angle$ EFG

ANSWERS: WORKSHEET 4-2

1. Obtuse
2. Acute
3. 103°
4. 77°
5. B
6. ∠ABD

ANSWERS: Chapter 1 Review Test

1. Obtuse
2. Acute
3. B = 130° C = 50° D = 130°
4. Adjacent
5. Adjacent
6. 129°
7. C = 51°
8. The arrows represent that lines don't end.
9. A Line segment.

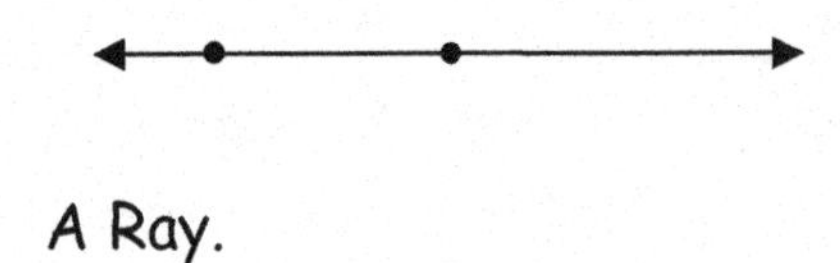

A Ray.

ANSWERS: Chapter 1 Review Test continued

9. An angle with a vertex Q.

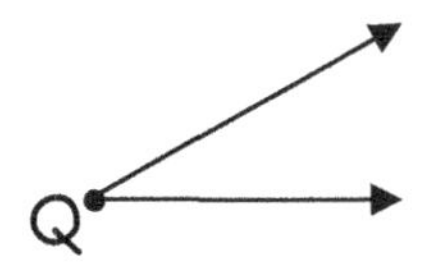

Any Angle.

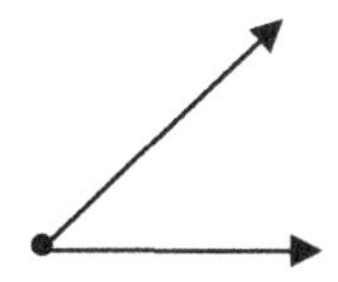

A Right angle.

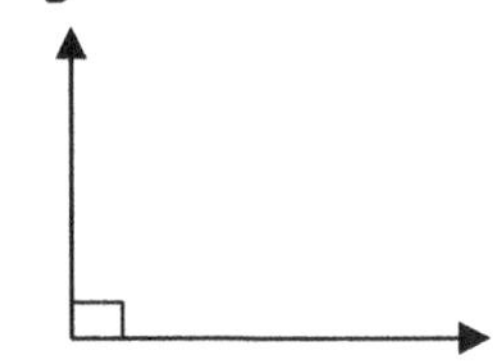

An Obtuse angle.

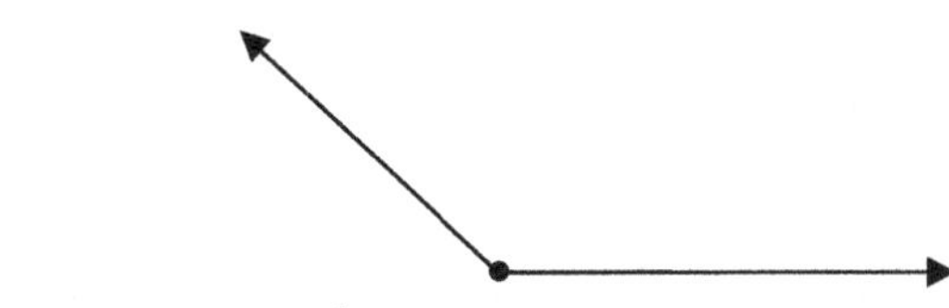

An Acute angle.

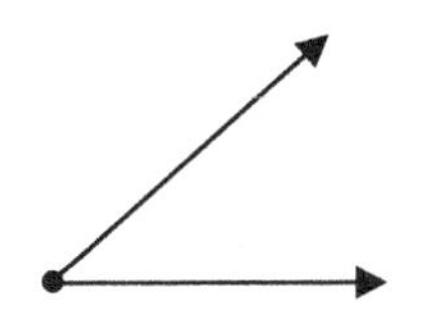

Two opposite angles.

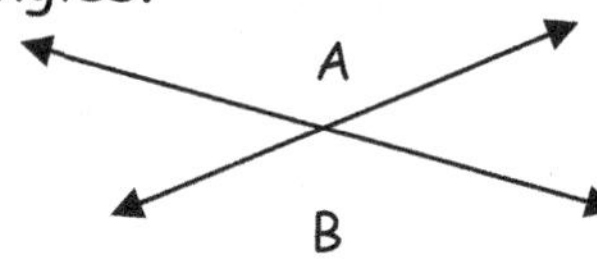

Two adjacent angles.

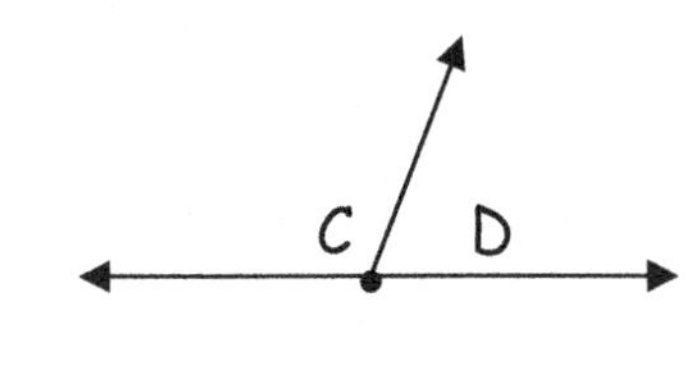

ANSWERS: WORKSHEET 4-4

1. a. Right Triangle
   b. Isosceles Triangle
   c. Equilateral

2. a. 40°
   b. 7"
   c. 50°

3. a. 60°
   b. 10'
   c. 180°

4 a. 90°
   b. 45°
   c. 40°

5. All triangles have 3 angles. If you measure all 3 angles and add them together what will be the total? **180°**

6. I'm thinking of a triangle. Two of the angles are 60°. What kind of triangle is it? **Equilateral**

7. I have a right triangle. One of the angles measures 50°. What are the measurements of the other two angles? **90° and 40°**

8. I have a triangle. The measurements of the angles are 30°, 60°, and 90°. What type of triangle do I have? **Right Triangle**

9. What does it take for a triangle to be called an equilateral triangle>
   All equal sides and all equal angles.
10. What does it take for a triangle to be called an isosceles triangle?
   At least two equal sides and angles.
11. Is it possible for one triangle to be all three special triangles? **No**

ANSWERS: WORKSHEET 4-5

1. Think of all the numbers between 1 and 20. Which number looks like two parallel lines? **Eleven 11**

2. Which two letters of the alphabet look like perpendicular lines? **T and L**

3. I have a shape with four 90° angles. Two of the four sides measure 2", the other two sides measure 4". What kind of shape do I have? **A Rectangle**

4. Which set of lines are parallel to each other? **A**

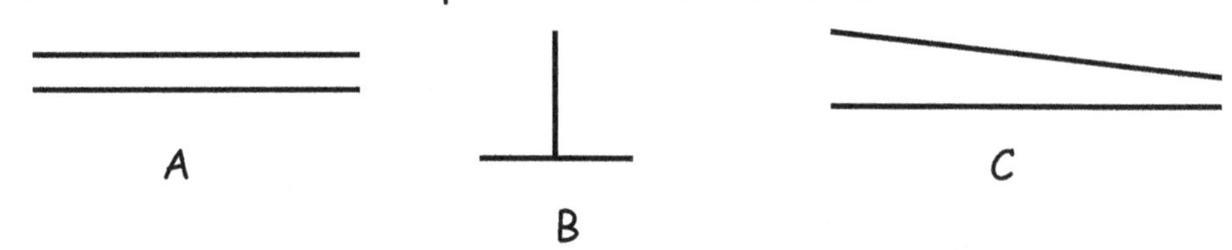

5. Look at A, B, and C above in problem number 4. Which set of lines are perpendicular to each other? **B**

6. Which angle is a 90 degree angle? **A**

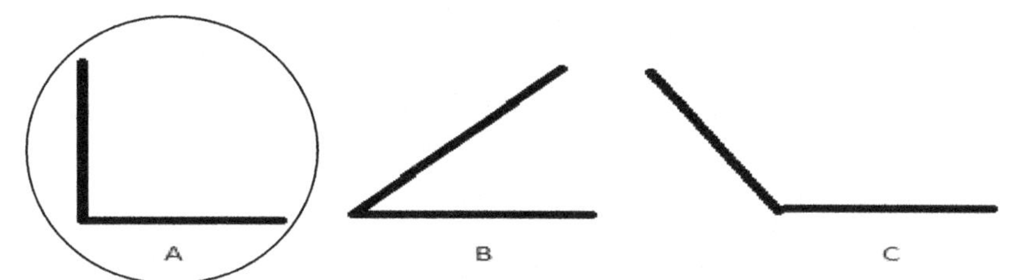

7. Which shape is a square? **B**

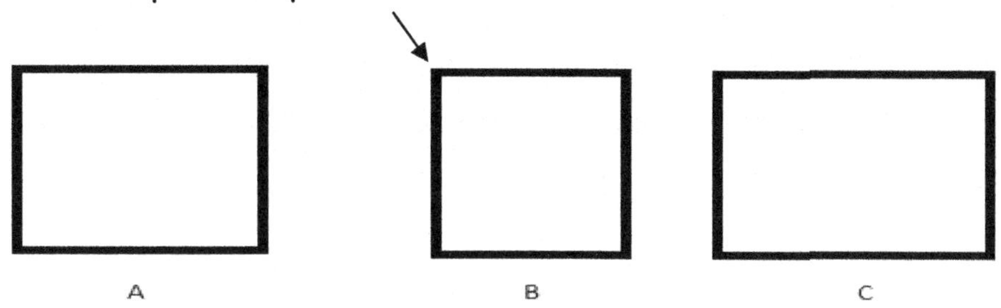

8. How many dimensions does a plane figure have? **2** - it's 2-Dimensional.

ANSWERS: WORKSHEET 4-6

1. Use A = bh to find the area of this square. **8" x 8" = 64 square inches.**

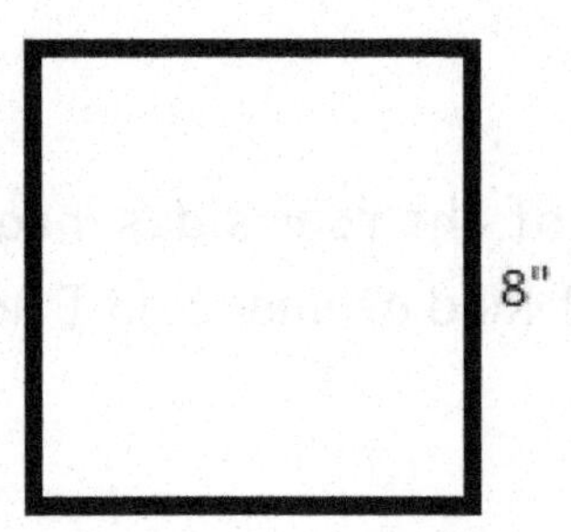

2. Find the area of this rectangle. **10 yards x 40 yards = 400 sq. yards.**

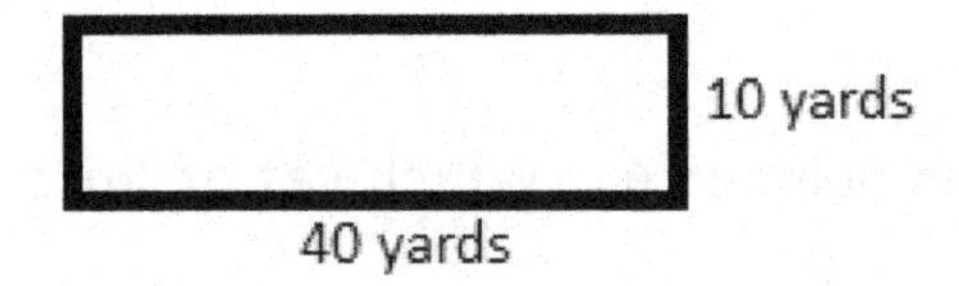

3. Your back yard measures 30 feet x 40 feet. What is the area of the back yard? **1,200 square feet**

4. What is the area of a 9 foot tall square? **9' x 9' = 81 square feet**

5. My Geometry book measures 11" x 8". How many square inches are on the cover? **11" x 8" = 88 square inches**

6. I'm going to put tile on a floor. Each tile is 1 square foot. The floor measures 20' on one side and 24' on the other side. How many tiles will I need to cover the floor? **20' x 24' = 480 sq. ft. I need 480 tiles.**

ANSWERS: WORKSHEET 4-7

Find the area of the following triangles using the formula $A = \frac{1}{2}bh$.

1. 10" 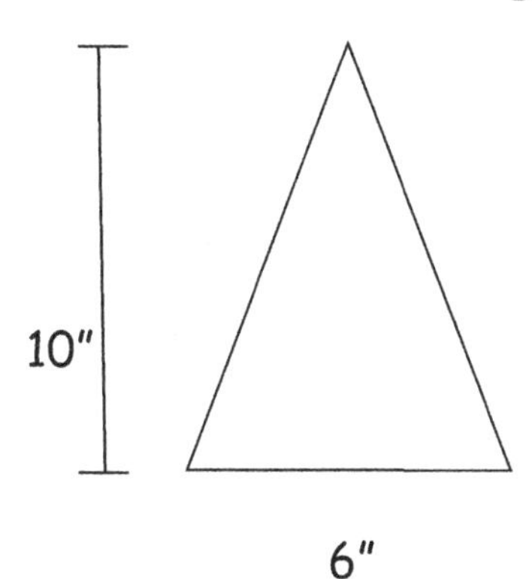 6"

$$A = \frac{1}{2}bh$$
$$A = \frac{1}{2}(10 \times 6)$$
$$A = \frac{1}{2}(60)$$
$$A = 30\ sq.inches$$

2. 9' 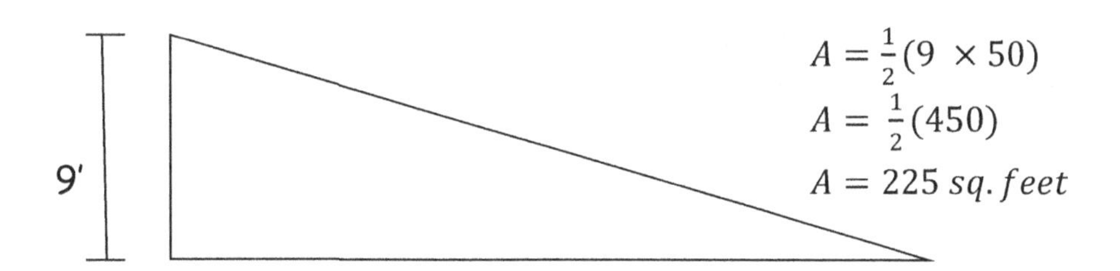 50'

$$A = \frac{1}{2}(9 \times 50)$$
$$A = \frac{1}{2}(450)$$
$$A = 225\ sq.feet$$

3. 3 miles 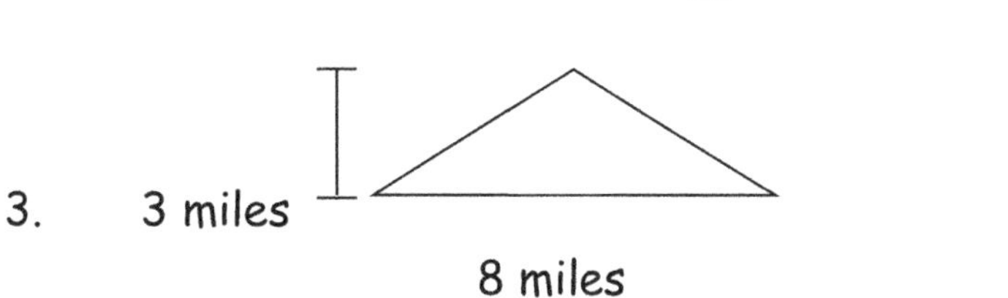 8 miles

$$A = \frac{1}{2}(3 \times 8)$$
$$A = \frac{1}{2}(24)$$
$$A = 12\ sq.miles$$

4. Look at triangle ABC and then answer the following questions. You won't need a protractor or a ruler. All you need to know is that it is an **equilateral** triangle and side AC = 4".

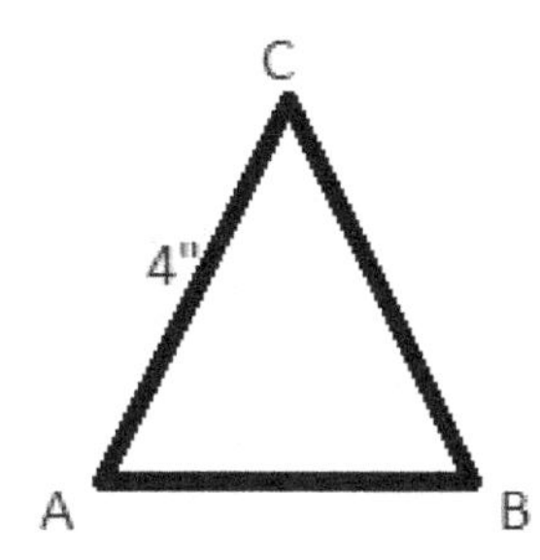

a. What is the length of side BC? **4"**

b. What is the size of angle A? **60°**

ANSWERS: WORKSHEET 4-7 continued

c. What is the length of side AB? **4"**

d. What is the size of angle C? **60°**

5. Look at triangle DEF and answer the following questions. You won't need a protractor or a ruler all you need to know is that it is a **right** triangle.

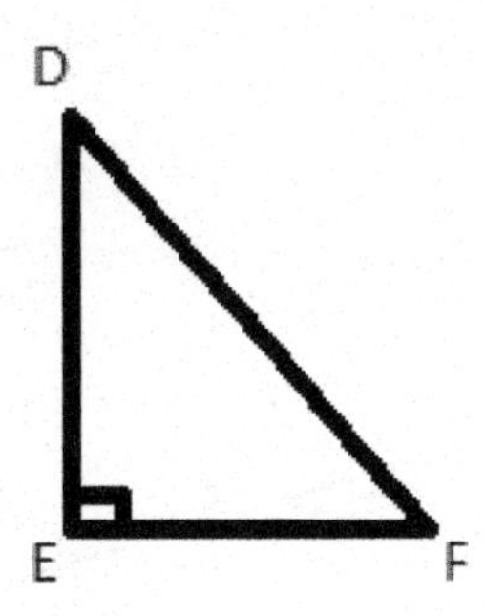

a. What size is angle E? **90°**

b. Angle D is 45°. What size is angle F? **45°**

c. What does the little box in angle E mean? **It means that angle is exactly 90°.**

6. I'm going to paint a triangle on the wall. The triangle will be 8 feet tall. The base of the triangle will be 5 feet. How many square feet of wall will I be painting?

8'

5'

$A = \frac{1}{2}(8 \times 5)$

$\boldsymbol{A = 20\ sq.ft.}$

7. If I fold any square in half diagonally, what two new shapes will I get?

I will create 2 triangles.

ANSWERS: WORKSHEET 4-8

Name each polygon.

1.

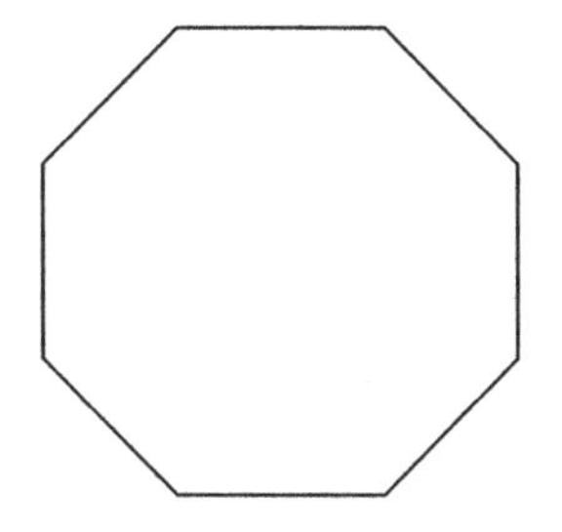

Octagon

2.

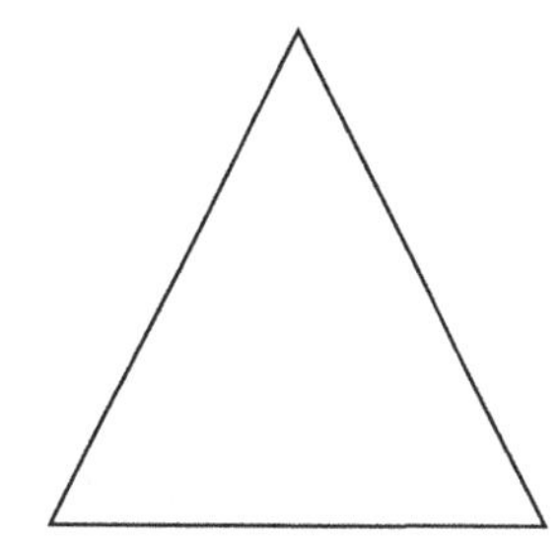

Triangle

3.

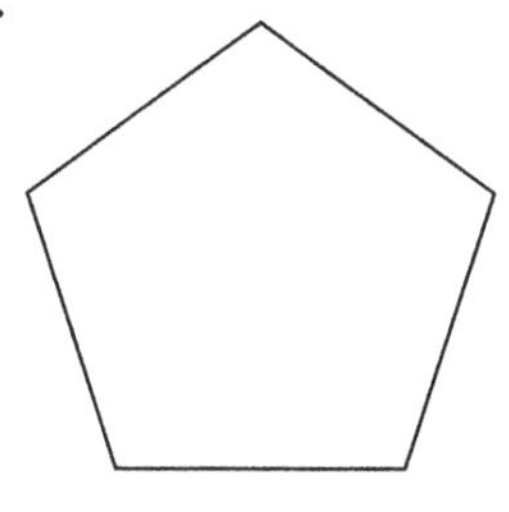

Pentagon

4.

Rectangle or Quadrilateral

5.

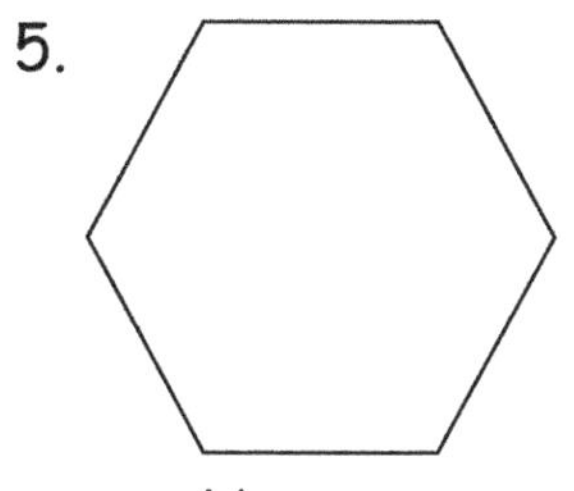

Hexagon

6.

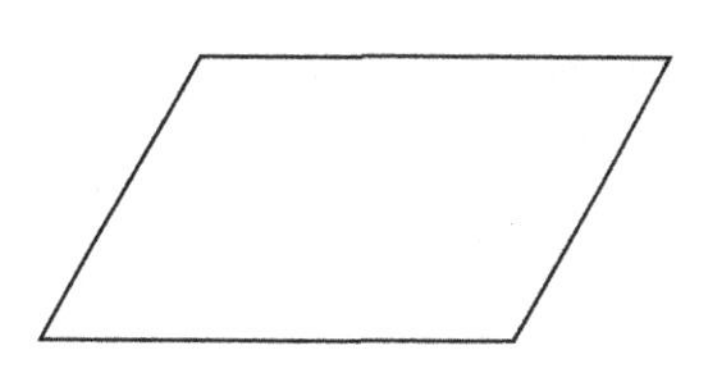

Quadrilateral

7.

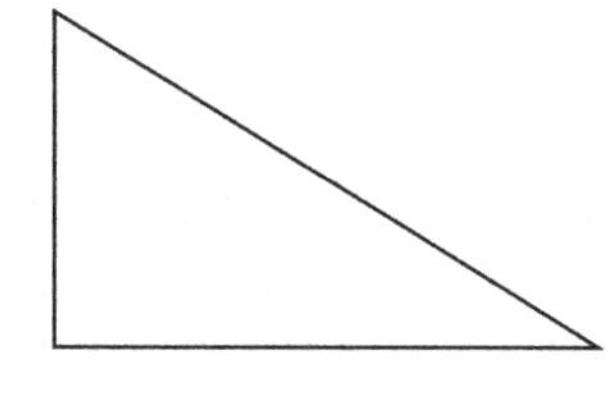

Triangle

8.

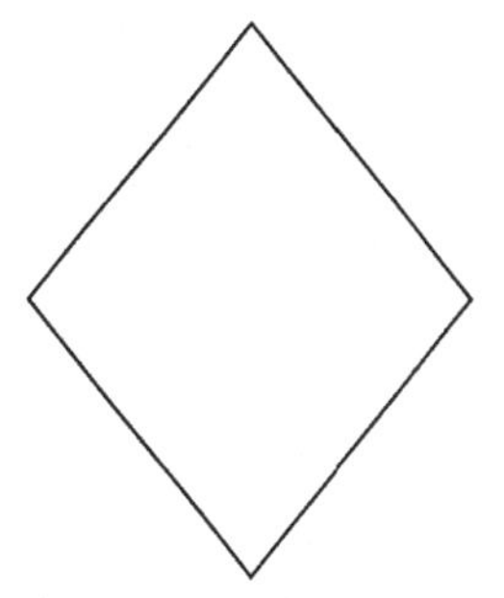

Quadrilateral

ANSWERS: Chapter 2 Review Test

Name the following shapes:

1. **Pentagon** 2. **Rectangle or quadrilateral** 3. **Octagon**

Find the area of the following shapes:

4. 7" x 14" = 98 in^2
   98 in^2 x $\frac{1}{2}$ = **49 square inches**

5. 6' x 18 ' = **108 square feet**

6. 12" x 10" = 120in^2
   120 in^2 x $\frac{1}{2}$ = **60 square inches**

7. 11' x 11' = **121 square feet**

8. Name the 3 special triangles.
   Isosceles **Right Triangle** **Equilateral**

9. Which of the 3 special triangles has 60° angles on all 3 sides?
   Equilateral Triangle

10. Which of the 3 special triangles has at least 2 equal sides?
    Isosceles Triangle

11. Which of the 3 special triangles has a 90° angle.
    Right Triangle

12. What is the total of all three angles in an isosceles triangle?
    180°

13. Brianna drew a plane figure on a piece of paper. The figure has 4 angles with opposite sides being parallel. The top and bottom are both 6" long. The 2 sides are both 8" long. What type of shape did Brianna draw?
    Brianna drew a rectangle, or a quadrilateral.

ANSWERS: WORKSHEET 4-9

1. Use the Pythagorean Theorem to find the length of this hypotenuse.

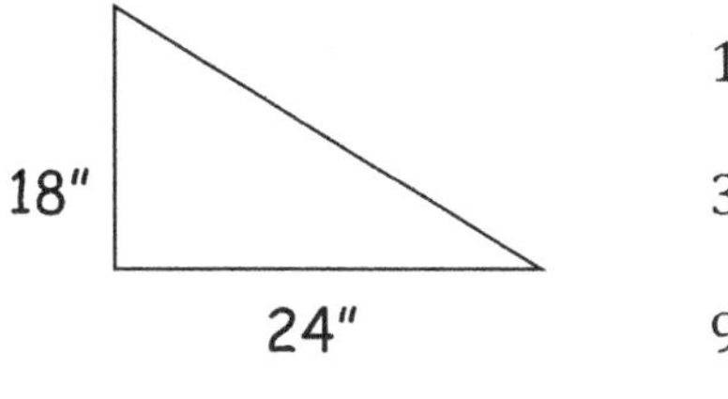

$$18^2 + 24^2 = c^2$$

$$324 + 576 = c^2$$

$$900 = c^2$$

$$\sqrt{900} = \sqrt{c^2}$$

$30 = c$ *the hypotenuse is 30" long*

2. What type of triangle has a hypotenuse? **Right Triangle**

3. Triangle ABC is an isosceles triangle. What is the length of side AB?

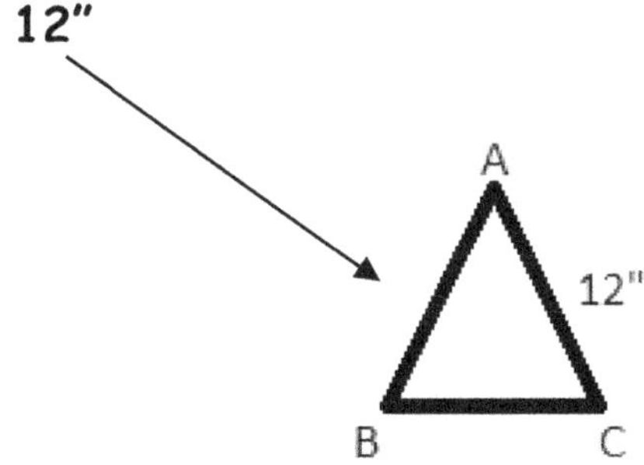

4. Find the length of the hypotenuse for triangles A, B, and C.

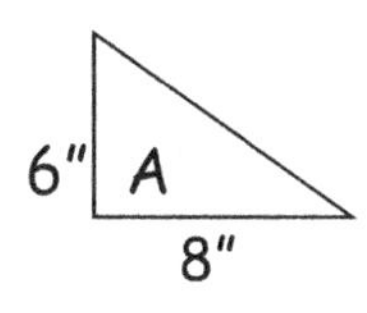

$6^2 + 8^2 = c^2$

$36 + 64 = c^2$

$100 = c^2$ $10" = c$

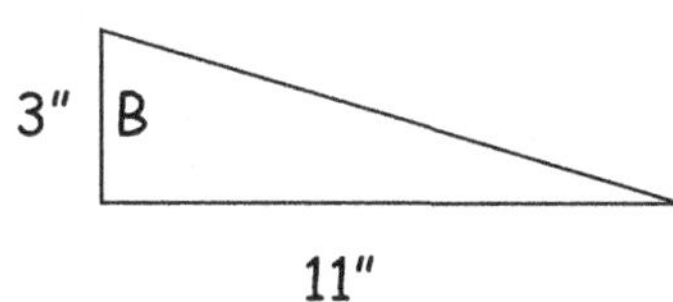

$3^2 + 11^2 = c^2$

$9 + 121 = c^2$

$130 = c^2$ $\sqrt{130}" = c$

10' C

4'

$10^2 + 4^2 = c^2$

$100 + 16 = c^2$

$116 = c^2$ $\sqrt{116}' = c$

ANSWERS: Chapter 3 Review Test

1. The hypotenuse is 5 feet because it is a 3-4-5 triangle.

2. Find the length of the hypotenuse.

$$12^2 + 16^2 = c^2$$
$$144 + 256 = c^2$$
$$400 = c^2$$
$$20 = c$$

3. Find the diagonal length of this rectangle. **18 is divisible by 3 and 24 is divisible by 4. That makes this right triangle a 3-4-5 triangle. Each side is multiplied by 6. Do the math. 5 x 6 = 30, the hypotenuse is 30".**

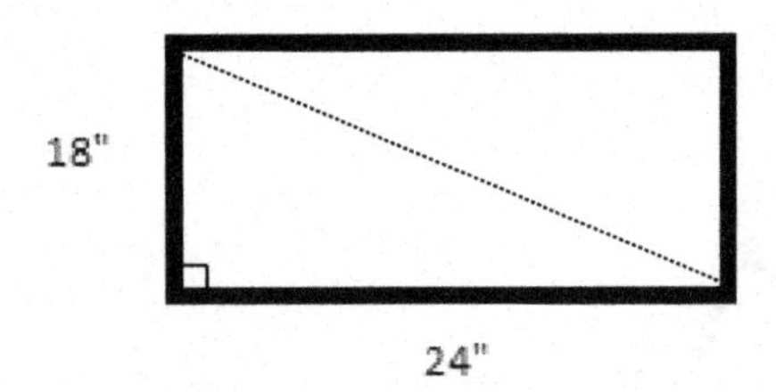

4. The rectangle for the football game is 100 yards long by 25 yards wide. What is the diagonal measurement of the football field?

$$25^2 + 100^2 = c^2$$
$$625 + 10{,}000 = c^2$$
$$10{,}625 = c^2$$
$$\sqrt{10625}\ yards = c \ \ or \ \ 103.077\ yards = c$$

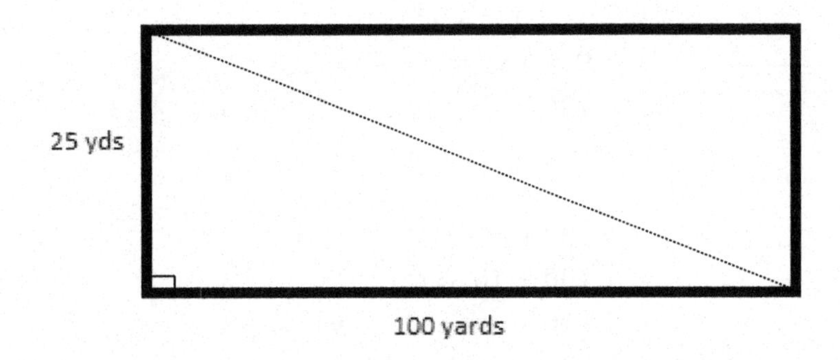

ANSWERS: Chapter 3 Review Test page 2

5. Write the Pythagorean Theorem. $a^2 + b^2 = c^2$
6. What is the missing measurement? **4 x 9 = 36 x = 36**

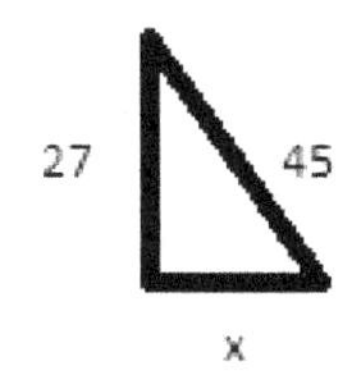

Compare this triangle to a 3-4-5 triangle.

Side a is 27 and side c is 45.

27 ÷ 3 = 9 and

45 ÷ 5 = 9

So side b or "4" is also multiplied by 9. The missing side is 36.

You could also solve this by using the Pythagorean Theorem.

$27^2 + x^2 = 45^2$

$729 + x^2 = 2{,}025$

$x^2 = 2{,}025 - 729$

$x^2 = 1{,}296$

$\sqrt{x^2} = \sqrt{1296}$

x = 36

ANSWERS: WORKSHEET 4-11

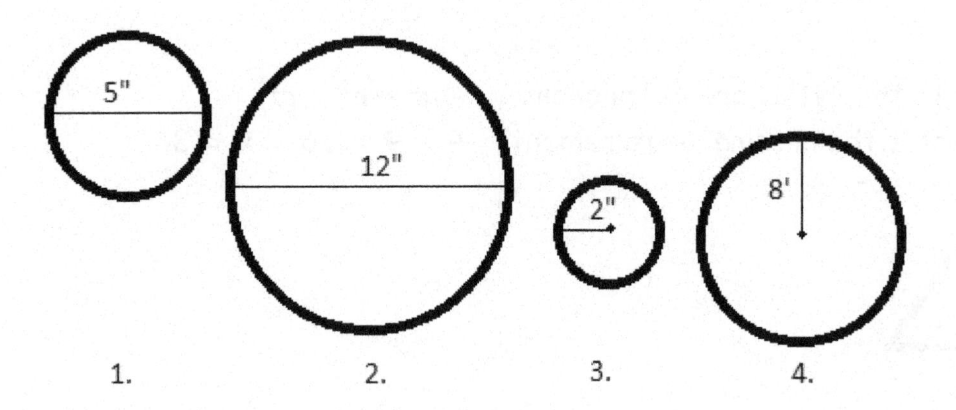

1. $c = 5" \times 3.14$
   $\boldsymbol{c = 15.7"}$

2. $c = 12" \times 3.14$
   $\boldsymbol{c = 37.68"}$

3. $c = 4" \times 3.14$
   $\boldsymbol{c = 12.56"}$

4. $c = 16' \times 3.14$
   $\boldsymbol{c = 50.24'}$

5. Which Greek symbol do we use to represent 3.14 inches? π pi

6. Mick cut down a tree with a chainsaw. The stump measured 6.28 feet around. What is the diameter of the tree he cut down? Automatically I know the answer is 2 feet in diameter because the diameter is twice as much as pi. You could also divide 6.28 by pi to get the answer. $\boldsymbol{6.28' \div 3.14 = 2'}$

7. Josh bought a round swimming pool. It measures 15′ from one side to the other side. Josh wants to put a string of Christmas lights around the pool. How long should the string of lights be? $c = 15' \times 3.14$
   $c = 47.1'$ The Christmas lights should be at least **47.1 feet long**.

ANSWERS: WORKSHEET 4-11 continued

8. Amanda drove over the bridge from point A to point B. It was a 1 mile drive across the bridge. The circle below represents a lake. How long would she have driven if she had gone around the lake from point A to point B? Hint: Half the circumference.

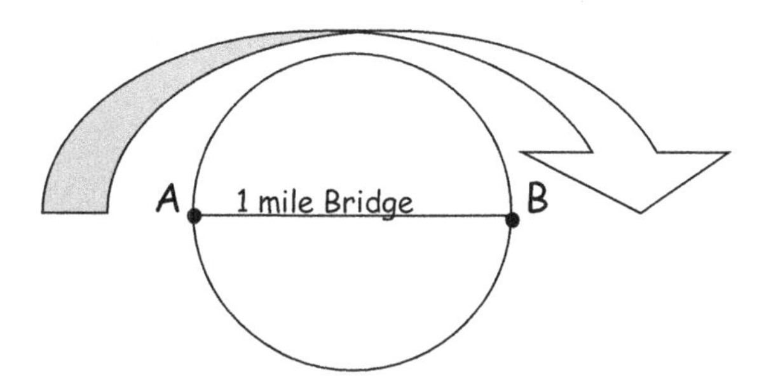

First, find the circumference of the entire circle.

$$c = 1 \times 3.14$$

$$c = 3.14\,miles$$

Amanda drove only half way around the circle, so cut that number in half.

$$3.14 \times .5 = \mathbf{1.57\,miles}$$

ANSWERS: WORKSHEET 4-12

1. Use $C = d\pi$ to find the circumference of a circle. The diameter is 3.

   $C = 3\pi$

   $C = 9.42$ inches (Not square inches, this is a length of string not a rug).

2. Use $A = \pi r^2$ to find the area of a circle. The diameter is 8" so the radius is 4".

   $A = \pi 4^2$

   $A = \pi 16$

   A = 50.24 square inches (You must say "square inches" when you are talking about area, it's in the formula).

3. You should have immediately known that the circumference would be **3.14 inches**. If you got this one wrong, you need to read the section on circles again.

4. If the circumference of a circle is 3.14 inches, then the diameter would have to be 1 inch. If the diameter is 1 inch, then the radius would be half that or **.5 inch or $\frac{1}{2}$ inch**.

ANSWERS: WORKSHEET 4-13

1. What is the area of this shape?

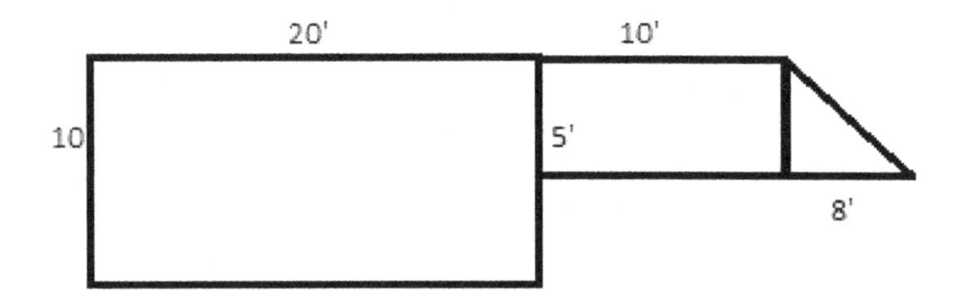

Large rectangle:	10′ x 20′ =	200 sq. ft.
Small rectangle:	10′ x 5′ =	50 sq ft.
Triangle:	8′ x 5′ = 40 x ½ =	20 sq. ft.
TOTAL:		**270 sq ft**

2. What is the area of this shape?

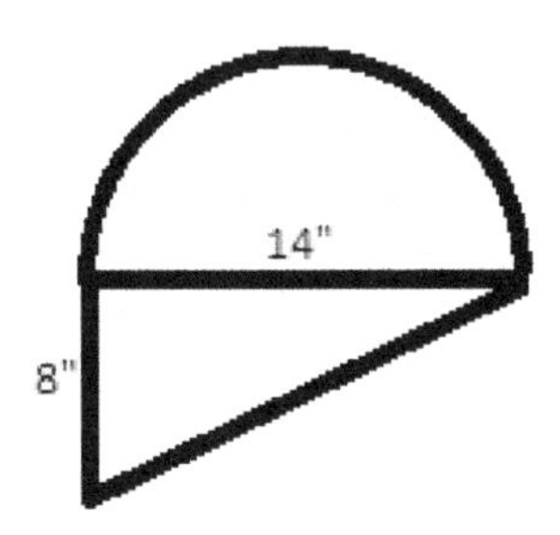

First, solve the triangle $A = \frac{1}{2}bh$
8 x 14 = 112 112 x ½ = 56 square inches
Next, is the circle $A = \pi r^2$
The diameter is 14, so the radius is 7 7 x 7 = 49
A = 3.14 x 49 = 153.86 square inches.
Since it is only a half circle, cut that number in half.
153.86 ÷ 2 = 76.93 square inches
Add the area of the triangle to the area of the half circle.
76.93 + 56 = 132.93 square inches

ANSWERS: WORKSHEET 4-13 continued

3. Find the area of the gray and white shapes below.

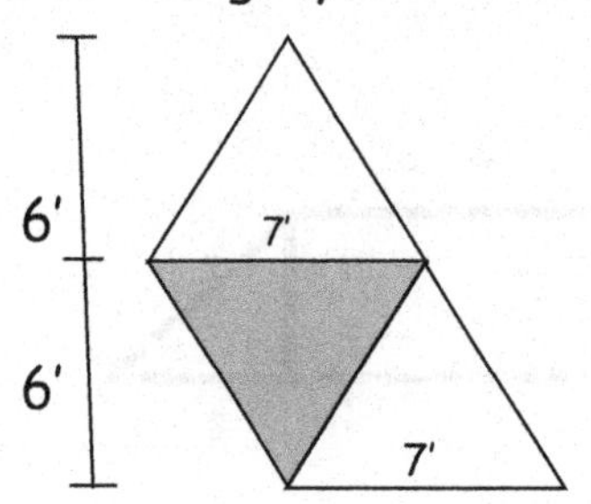

Each triangle has a base of 7' and a height of 6'. Use $A = \frac{1}{2}bh$.

Top triangle:	7 x 6 = 42	42 x $\frac{1}{2}$ = 21 sq. ft.
Gray triangle:	7 x 6 = 42	42 x $\frac{1}{2}$ = 21 sq. ft.
Bottom triangle:	7 x 6 = 42	42 x $\frac{1}{2}$ = 21 sq. ft.
	TOTAL	**63 square feet**

ANSWERS: WORKSHEET 4-14

1. Find the **perimeter** of the following three shapes: a square, a right triangle and a rectangle.

   A. This shape is a square, so all 4 sides are equal.

   5″ + 5″ + 5″ + 5″ = **20″**.

   B. This shape is a right triangle. It is a 3-4-5 triangle, so the perimeter is:

   3′ + 4′ + 5′ = **12 feet**.

   C. This shape is a rectangle, 12′ + 12′ + 20′ + 20′ = **64 feet**.

2. C = dπ

   The radius is 5″, so the diameter is 10″.

   C = 10″ x 3.14

   C = 31.40″

   The circumference of the circle is 31.4 inches.

3. Find the perimeter of a square with 3/8″ sides.

   $\frac{3}{8} \times \frac{4}{1} = \frac{12}{8} = 1\frac{4}{8} = \mathbf{1\frac{1}{2}} \; \boldsymbol{inches}$

ANSWERS: WORKSHEET 4-15

1. 12' x 6' x 8' = 576 cubic feet
2. 5" x 5" x 25" = 625 inches^3
3. 4' x 2' x 1' = 8ft^3

ANSWERS: WORKSHEET 4-16

1. A basketball with a radius of 7″.

$$v = \frac{4}{3}\pi(7^3)$$
$$v = \frac{4}{3} \times \frac{314}{100} \times \frac{343}{1} = \frac{430808}{300}$$
$$v = 1{,}436\frac{8}{300} = 1{,}436\frac{2}{75} in^3$$

2. A globe with a radius of 10″.

$$v = \frac{4}{3}\pi(10^3)$$
$$v = \frac{4}{3} \times \frac{314}{100} \times \frac{1000}{1}$$
$$v = \frac{1{,}256{,}000}{300}$$
$$v = 4186\frac{200}{300} = 4{,}186\frac{2}{3} in^3$$

3. A bouncy ball with a radius of ½ inch.

$$v = \frac{4}{3}\pi\left(\frac{1}{2}^3\right)$$
$$v = 1.33 \times 3.14 \times 0.125$$
$$v = .52 in^3$$

Find the volume of the following rectangular prisms.

4. A box that has a base of 4″, stands 5″ tall, and has a width of 7″.
   $v = 4" \times 5" \times 7"$ $\quad$ $\boldsymbol{v = 140\ cubic\ inches}$

5. A treasure chest that is 3′ wide, 2′ tall, and 2′ deep.
   $v = 3' \times 2' \times 2'$ $\quad$ $\boldsymbol{v = 12\ cubic\ feet}$

6. A cube of butter that measures 1″ tall, 2.25″ long, and 1-1/8″ wide.
   $v = 1" \times 2.25" \times 1.125"$ $\quad$ $v = 2.53\ in^3$

7. What is a space figure? **A 3-Dimensional shape.**
8. What is the difference between a cube and a square?
   A cube is a space figure with 3 dimensions. A square is a plane figure with only 2 dimensions.
9. What is the difference between a circle and a sphere?
   A circle is a flat plane figure. A sphere is a 3-D space figure.
10. What is the difference between a cube and a rectangular prism?
   A cube has all equal sides. A rectangular prism can have 3 different dimensions.

ANSWERS: Chapter 4 Review Test

Find the **C**ircumference of the following circles.

1. A circle with a diameter of 10 yards.

$C = d\pi$
$C = 10 \times 3.14$
$C = 31.40\ yards$

2. A circle with a radius of 7 miles.

$C = 14 \times 3.14$
$C = 43.96\ miles$

Find the **A**rea of the following circles.

3. A circle with a radius of 3 feet.

$A = \pi r^2$
$A = 3.14 \times 3^2$
$A = 28.26\ sq.ft.$

4. A circle with a diameter of 12 inches.

$A = 3.14 \times 36\ inches$
$A = 113.04\ sq.in.$

Find the area of the following shapes.

5.

15 feet

9 feet

12 feet

$Rectangle\text{: } 9\ feet \times 15\ feet = 135\ sq.ft.$
$Triangle\text{: } 9 \times 12 = 108 \quad 108 \times \frac{1}{2} = 54\ sq.ft.$
$135 + 54 = 189\ sq.ft.$

6. Find the perimeter of the shape above.

$Find\ the\ hypotenuse\ a^2 + b^2 = c^2$
$81 + 144 = 225 \qquad \sqrt{225} = 15$
$Add\ the\ sides\text{: } 9' + 15' + 15' + 12' + 15' = 66\ feet$

ANSWERS: Chapter 4 Review Test page 2

Find the volume of the following space figures.

7.

$6' \times 20' \times 5' = \mathbf{600\ ft^3}$

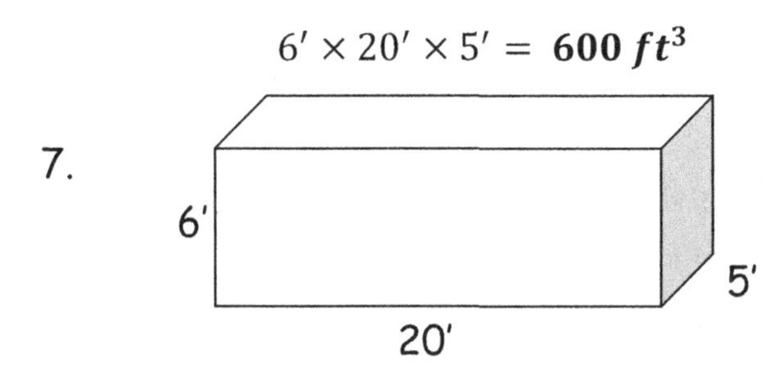

8.

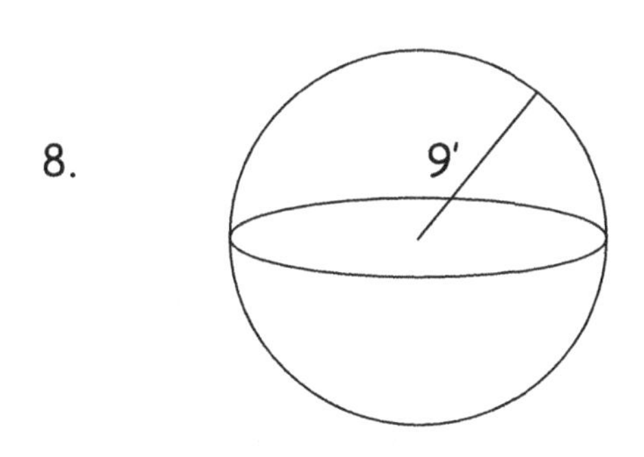

$Vol.\ of\ sphere = \frac{4}{3}\pi r^3$

$V = \frac{4}{3} \times 3.14 \times 9^3$

$\frac{4}{3} \times \frac{314}{100} \times \frac{729}{1} =$

$\frac{915{,}624}{300} = 3{,}052\frac{24}{300} = \mathbf{3{,}052\frac{2}{25}ft^3}$

ANSWERS: WORKSHEET 4-17

Fill in the blanks with the appropriate unit of measurement. Answer with meters, liters, or grams.

1. The golf ball traveled 200 **METERS**.
2. The letter only weighs 20 **GRAMS**, so 1 stamp is enough postage.
3. A bottle of water measures 1/2 **LITER**.
4. How many **GRAMS** does a penny weigh?
5. How many **LITERS** of gas did you put in your car?
6. It took me almost an hour to walk 3000 **METERS**.
7. Read the back of the box. How many **GRAMS** of sugar are in this snack?
8. A bag of chips holds 340 **GRAMS** of chips.
9. The fish tank broke and spilt **LITERS** and **LITERS** of water on the floor.
10. Did you see Mick jump? He must have traveled at least 3 **METERS**.

ANSWERS: WORKSHEET 4-18

1. One paper clip weighs about 1 gram. If I break that paper clip into 1000 equal pieces, each piece will weigh 1 **milligram.**

2. I bought 1 liter of pop. I gave 1000 people a few drops each. Each person received 1 **milliliter** of pop.

3. I'm trying to measure how thin my credit card is. It is hard to measure something so small, but I know if I stack up 1000 credit cards, it will stand 1 meter tall. What is the thickness of just 1 credit card?
   1 millimeter

4. Sarah weighed herself. She weighed over 200 **kilograms.**

5. Jack filled up the swimming pool. It took over **3 *kiloliters*** of water to fill it.

6. The pharmacist said these pills have 100 **milligrams** of vitamin C.

7. I stacked up 10 dimes. The whole stack measured 10 **millimeters** tall.

8. The speed limit on the highway is 80 **kilometers** per hour in Canada.

9. The 10-K race was 10 **kilometers** long.

10. The recipe calls for 5 **milliliters** of oil.

11. That bottle has 120 **milliliters** of perfume in it.

12. 1 gram x 1000 = 1 **kilogram.**

13. 1 gram ÷ 1000 = **1 milligram**.

ANSWERS: WORKSHEET 4-19

Use one of the words from the list below to complete these sentences.

Kilogram	= 1000 grams	Centigram	= 1/100 of a gram
Kilometer	= 1000 meters	Centimeter	= 1/100 of a meter
Kiloliter	= 1000 liters	Centiliter	= 1/100 of a liter

Milligram = 1/1000 of a gram
Millimeter = 1/1000 of a meter
Milliliter = 1/1000 of a liter

1. I got a splinter in my hand. It was about 1 **millimeter** wide.

2. I drove to the store. It was about 3 **kilometers** away.

3. That fish tank is huge. It must have nearly 100 **liters** of water in it. **Or it could be 100 Kiloliters, but that is super huge.**

4. My doctor said this pill has 500 **milligrams** of aspirin in it.

5. A small bird feather weighs approximately 7**centigrams**.

6. I went on a diet and I lost over 9 **kilograms**.

7. I put 1 spoonful of water in the dough, that's about 1 **centiliter**.

8. The diameter of a marble is about 1 **centimeter**.

9. A bottle of eye drops holds 30 milliliters, that's the same as 3 **centiliters.**

10. I ran 5 **kilometers** in 1 hour.

ANSWERS: WORKSHEET 4-20

1. m
2. cm
3. L

4 cL

5. kg
6. g
7. mg
8. km
9. mL
10. kL

ANSWERS: Chapter 5 Review Test

1. How many centimeters are in a decimeter? **10**
2. How many decimeters are in one meter? **10**
3. How many millimeters are in a centimeter? **10**
4. How many centimeters are in a meter? **100**
5. What measurement is closest to the width of a pencil lead?
   (1 mm) 1 cm 1 dm 1 m

6. What measurement is closest to the length of a new pencil?
   2 mm 2 cm (2 dm) 2 m

7. What measurement is closest to the height of a door?
   2 mm 2 cm 2 dm (2 m)

8. What measurement is most likely the size of a chocolate chip cookie?
   7 mm (7 cm) 7 dm 7 m

9. What does the prefix Kilo mean? **1000**
10. What does the prefix Milli mean? **1/1000**
11. Which is most likely 1 mL?
    A glass of milk Volume of a fish tank (3 rain drops)

12. Which is heavier?
    3 grams 9 centigrams (1 kilogram)

13. In the United States, we use feet for our base unit to measure length. What is the base unit used in the Metric System to measure length?
    Meter
14. Your protractor has a 1 decimeter ruler on it. How many millimeters are in 1 decimeter? **100mm**

ANSWERS: Chapter 5 Review Test page 2

15. Write the abbreviations for the following metric units.

kilometer = km
Liter = L
Hectometer = hm
Dekagram = dag
Meter = m
Deciliter = dL
Centimeter = cm
Milligram = mg
Gram = g
Milliliter = mL

ANSWERS: FINAL GEOMETRY TEST

1. **Ray AB** or $\overrightarrow{AB}$
2. Angle XYZ is written $\angle$XYZ for short or $\angle$**Y** for the shortest way.
3. Angle EBA = **90°**.
4. Angle DBC is **obtuse**.
5. Angle FBC is **acute**.
6. Angle ABD = **65°**.
7. Angle **DBE** is adjacent to angle ABD.
8. Angle ABF = **135°**.
9. Angle **EBF** is adjacent to angle FBC.
10. Angle DBE would be **20°**.

11. Angle **DEB** is opposite angle AEC.
12. The vertex is **E**.
13. Angle **AED** is opposite angle CEB.
14. Angles **AEC and DEB** are adjacent to angle AED.
15. Angle DEB is opposite, so it also measures **15°**.
16. Angle CEB would measure **165°**.
17. **Isosceles triangle**.
18. **Right triangle**.
19. **Equilateral triangle**.
20. Angle A = **60°** because B = 90 and C = 30 and they must all equal 180 together.

21. If angle A = 40°, then angles B and C must total 140°. Angles B and C must be the same, since it is an isosceles triangle, so they are **70° each**.

22. Since it is an equilateral triangle, all of the angles are the same. So angle A must be **60°**.

23. To find the area of a triangle use the formula A = $\frac{1}{2}$ bh. The base is 6" and the height is 5". 6" x 5" = 30sq in. Half of 30 is 15, so the area of the triangle is **15 square inches**.

ANSWERS: Final Geometry Test continued

24. To find the area of a rectangle use A = bh. 72′ x 12′ = **864 square feet.**

25. Area = πr^2 Area = 3.14 x 7^2 Area = 3.14 x 49 **Area = 153.86 square inches.** Circumference = $d\pi$ C = 14 x 3.14 **C = 43.96 inches.**

26. Area = 3.14 x 5.5^2 Area = 3.14 x 30.25 **Area = 94.985 sq. ft.** Circumference = 11′ x 3.14 **Circumference = 34.54ft.**

27. The perimeter of shape A is 10 + 15 + 10 + 15 = **50 inches.**
Using the 3-4-5 triangle secret, I can see that the hypotenuse is 10.
The perimeter of shape B is 6 + 8 + 10 = **24 feet.**

28. C

29. A

30. C

31. $3^2 + 4^2 = c^2$
$9 + 16 = c^2$
$25 = c^2$
5″ = c
Or you could have used the 3-4-5 triangle secret, to see that the hypotenuse is 5″.

32. $a^2 + b^2 = c^2$ $9^2 + 12^2 = c^2$ $81 + 144 = c^2$ $225 = c^2$ **15′ = c**

33. The shape is a square, so all sides are 12″. To find the diagonal, we use the Pythagorean Theorem $a^2 + b^2 = c^2$.
a = 12 and b = 12
$12^2 + 12^2 = c^2$.
$144 + 144 = c^2$
$288 = c^2$ **c =** $\sqrt{288}$

ANSWERS: Final Geometry Test continued

34. The area of the square is 8" x 8" = 64 sq inches. Since there are 2 half circles with the same diameter, I will get the area for 1 whole circle. I will use $A = \pi r^2$ to find the area of the circle. The diameter is 8", so the radius is 4".
A = 3.14 x 4^2
Area = 3.14 x 16 Area = 50.24 square inches.

The area of the square is $64in^2$ and the area of the circle halves are $50.24in^2$. Add those two together $64in^2 + 50.24in^2 =$ **$114.24in^2$**.

35. A cube has all equal sides, so the dimensions are 5" x 5" x 5". Use the volume formula V = bhw. Volume = 5" x 5" x 5" = **$125in^3$**.

36. Use the volume formula. V = bhw Volume = 20' x 4' x 3' = **$240ft^3$**.

37. This angle measures **46°**.

38. Pythagoras is on the cover of this book.

39. Volume = 4/3 πr^3

4/3 x 3.14 x 7^3

7" x 7" x 7" = $343in^3$

$\frac{4}{3} \times \frac{3.14}{1} \times \frac{343in^3}{1} = \frac{4308.08in^3}{3}$

$\frac{4308.08}{3}$ in^3 = **1436.03 in^3**

40. **1 kilometer.**
41. There are **1000 milliliters** in 1 liter.

ANSWERS: Final Geometry Test continued

42. The shape below has an area of 54cm^2. What is the length of base x?

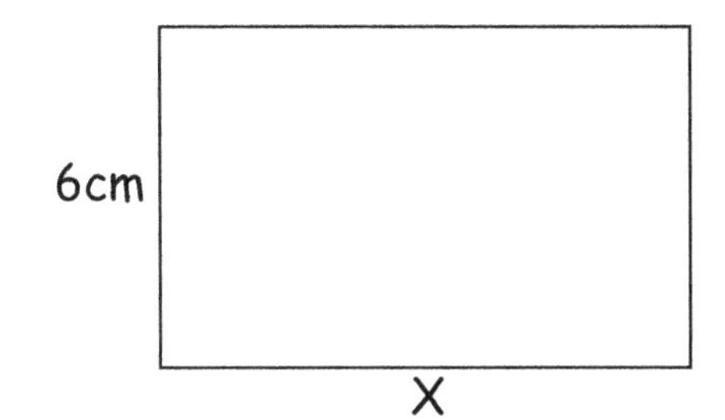

$54cm^2 \div 6cm = 9cm$

$\boldsymbol{x = 9cm}$

43. How many dimensions are on a plane figure? **2 dimensions**

44. How many dimensions are on a space figure? **3 dimensions**

45. Which metric unit of measurement is closest to the US Standard inch?

A centimeter

46. Which metric unit of measurement is closest to the US Standard teaspoon? **A centiliter**

47. Which metric unit of measurement is closest to the US Standard yard?

1 meter

48. Write the Pythagorean Theorem. $\boldsymbol{a^2 + b^2 = c^2}$

49. Which side of the triangle below is the hypotenuse?

Z is the hypotenuse

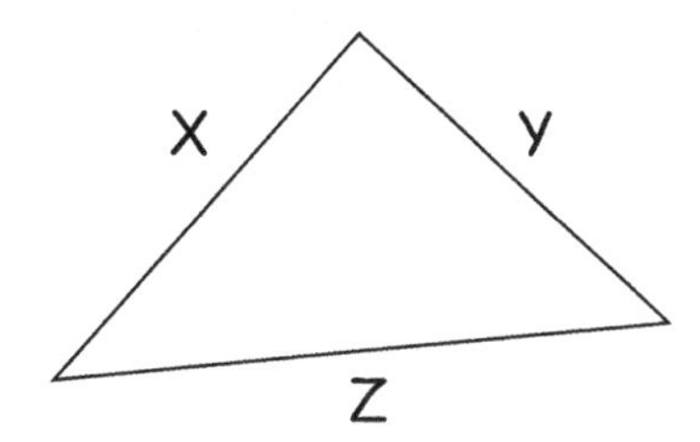

ANSWERS: Final Geometry Test continued

50. What is the name of a 3-sided polygon? **A Triangle**

51. What is the name of a 5-sided polygon? **A Pentagon**

52. Give 3 names for a 4-sided polygon. **Square, Rectangle, Quadrilateral**

53. How many sides does an octagon have? **8 sides**

54. How many sides does a hexagon have? **6 sides**

55. Is a circle a polygon? **No, the edges are not straight.**

56. What formula has the nickname Seedy Pie? What do you use this formula for? $c = d\pi$ **is used to find the circumference of a circle.**

57. What formula has the nickname A Pie Are Squared? What do you use this formula for? $A = \pi r^2$ **is used to find the area of a circle.**

Made in the USA
Coppell, TX
11 September 2021